Data Analysis and Visualization Using Python

Analyze Data to Create Visualizations for BI Systems

Dr. Ossama Embarak

Apress®

Data Analysis and Visualization Using Python

Dr. Ossama Embarak
Higher Colleges of Technology, Abu Dhabi, United Arab Emirates

ISBN-13 (pbk): 978-1-4842-4108-0 ISBN-13 (electronic): 978-1-4842-4109-7
https://doi.org/10.1007/978-1-4842-4109-7

Library of Congress Control Number: 2018964118

Managing Director, Apress Media LLC: Welmoed Spahr
Acquisitions Editor: Nikhil Karkal
Development Editor: Matthew Moodie
Coordinating Editor: Divya Modi

Cover designed by eStudioCalamar

Cover image designed by Freepik (www.freepik.com)

Distributed to the book trade worldwide by Springer Science+Business Media New York, 233 Spring Street, 6th Floor, New York, NY 10013. Phone 1-800-SPRINGER, fax (201) 348-4505, e-mail orders-ny@springer-sbm.com, or visit www.springeronline.com. Apress Media, LLC is a California LLC and the sole member (owner) is Springer Science + Business Media Finance Inc (SSBM Finance Inc). SSBM Finance Inc is a **Delaware** corporation.

For information on translations, please e-mail rights@apress.com, or visit www.apress.com/rights-permissions.

Apress titles may be purchased in bulk for academic, corporate, or promotional use. eBook versions and licenses are also available for most titles. For more information, reference our Print and eBook Bulk Sales web page at www.apress.com/bulk-sales.

Any source code or other supplementary material referenced by the author in this book is available to readers on GitHub via the book's product page, located at www.apress.com/978-1-4842-4108-0. For more detailed information, please visit www.apress.com/source-code.

Printed on acid-free paper

This book is dedicated to my family—my mother,
my father, and all my brothers—for their endless support.

Table of Contents

About the Author

Dr. Ossama Embarak holds a PhD in computer and information science from Heriot-Watt University in Scotland, UK. He has more than two decades of research and teaching experience with a number of programming languages including C++, Java, C#, R, Python, etc. He is currently the lead CIS program coordinator for Higher Colleges of Technology, UAE's largest applied higher educational institution, with more than 23,000 students attending campuses throughout the region.

Recently, he received an interdisciplinary research grant of 199,000 to implement a machine learning system for mining students' knowledge and skills.

He has participated in many scholarly activities as a reviewer and editor for journals in the fields of computer and information science including artificial intelligence, data mining, machine learning, mobile and web technologies. He supervised a large number of graduation projects, as well as he has published numerous papers about data mining, users online privacy, semantic web structure and knowledge discovery. Also he participated as a co-chair for numerous regional and international conferences.

About the Technical Reviewers

Shankar Rao Pandala is a data scientist at Cognizant. He has a bachelor's degree in computer science and a master's degree in financial markets. His work experience spans finance, healthcare, manufacturing, and consulting. His area of interest is artificial intelligence for trading.

Prashant Sahu has a bachelor's of technology from NIT Rourkela (2003) and is currently pursuing a doctorate from the Indian Institute of Technology, Bombay, in the area of instrumentation, data analytics, modeling, and simulation applied to semiconductor materials and devices. He is currently the head of training services at Tech Smart Systems in Pune, India. He is also mentoring the startup Bharati Robotic Systems (India) as an SVP of innovation. He has more than 15 years of experience in research, automation, simulation and modeling, data analytics, image processing, control systems, optimization algorithms, genetic algorithms, cryptography, and more, and he has handled many projects in these areas from academia and industry. He has conducted several faculty development training programs across India and has conducted corporate training for software companies across India. He is also an external examiner for B.E./M.E. projects and a member of the Syllabus Revision Committee at the University of Pune.

Introduction

This book looks at Python from a data science point of view and teaches the reader proven techniques of data visualization that are used to make critical business decisions. Starting with an introduction to data science using Python, the book then covers the Python environment and gets you acquainted with editors like Jupyter Notebooks and the Spyder IDE. After going through a primer on Python programming, you will grasp the fundamental Python programming techniques used in data science. Moving on to data visualization, you will learn how it caters to modern business needs and is key to decision-making. You will also take a look at some popular data visualization libraries in Python. Shifting focus to collecting data, you will learn about the various aspects of data collections from a data science perspective and also take a look at Python's data collection structures. You will then learn about file I/O processing and regular expressions in Python, followed by techniques to gather and clean data. Moving on to exploring and analyzing data, you will look at the various data structures in Python. Then, you will take a deep dive into data visualization techniques, going through a number of plotting systems in Python. In conclusion, you will go through two detailed case studies, where you'll get a chance to revisit the concepts you've grasped so far.

This book is for people who want to learn Python for the data science field in order to become data scientists. No specific programming prerequisites are required besides having basic programming knowledge.

Specifically, the following list highlights what is covered in the book:

- Chapter 1 introduces the main concepts of data science and its life cycle. It also demonstrates the importance of Python programming and its main libraries for data science processing. You will learn how different Python data structures are used in data science applications. You will learn how to implement an abstract series and a data frame as a main Python data structure. You will learn how to apply basic Python programming techniques for data cleaning and manipulation. You will learn how to run the basic inferential statistical analyses. In addition, exercises with model answers are given for practicing real-life scenarios.
- Chapter 2 demonstrates how to implement data visualization in modern business. You will learn how to recognize the role of data visualization in decision-making and how to load and use important Python libraries for data visualization. In addition, exercises with model answers are given for practicing real-life scenarios.
- Chapter 3 illustrates data collection structures in Python and their implementations. You will learn how to identify different forms of collection in Python. You will learn how to create lists and how to manipulate list content. You will learn about the purpose of creating a dictionary as a data container and its manipulations. You will learn how to maintain data in a tuple form and what the differences are between tuple structures and dictionary structures, as well as the basic tuples operations. You will learn how to create a series from

other data collection forms. You will learn how to create a data frame from different data collection structures and from another data frame. You will learn how to create a panel as a 3D data collection from a series or data frame. In addition, exercises with model answers are given for practicing real-life scenarios.

- Chapter 4 shows how to read and send data to users, read and pull data stored in historical files, and open files for reading, writing, or for both. You will learn how to access file attributes and manipulate sessions. You will learn how to read data from users and apply casting. You will learn how to apply regular expressions to extract data, use regular expression alternatives, and use anchors and repetition expressions for data extractions as well. In addition, exercises with model answers are given for practicing real-life scenarios.
- Chapter 5 covers data gathering and cleaning to have reliable data for analysis. You will learn how to apply data cleaning techniques to handle missing values. You will learn how to read CSV data format offline or pull it directly from online clouds. You will learn how to merge and integrate data from different sources. You will learn how to read and extract data from the JSON, HTML, and XML formats. In addition, exercises with model answers are given for practicing real-life scenarios.
- Chapter 6 shows how to use Python scripts to explore and analyze data in different collection structures. You will learn how to implement Python techniques to explore and analyze a series of data, create a series,

access data from a series with a position, and apply statistical methods on a series. You will learn how to explore and analyze data in a data frame, create a data frame, and update and access data in a data frame structure. You will learn how to manipulate data in a data frame such as including columns, selecting rows, adding, or deleting data, and applying statistical operations on a data frame. You will learn how to apply statistical methods on a panel data structure to explore and analyze stored data. You will learn how to statistically analyze grouped data, iterate through groups, and apply aggregations, transformations, and filtration techniques. In addition, exercises with model answers are given for practicing real-life scenarios.

- Chapter 7 shows how to visualize data from different collection structures. You will learn how to plot data from a series, a data frame, or a panel using Python plotting tools such as line plots, bar plots, pie charts, box plots, histograms, and scatter plots. You will learn how to implement the Seaborn plotting system using strip plots, box plots, swarm plots, and joint plots. You will learn how to implement Matplotlib plotting using line plots, bar charts, histograms, scatter plots, stack plots, and pie charts. In addition, exercises with model answers are given for practicing real-life scenarios.
- Chapter 8 investigates two real-life case studies, starting with data gathering and moving through cleaning, data exploring, data analysis, and visualizing. Finally, you'll learn how to discuss the study findings and provide recommendations for decision-makers.

CHAPTER 1

Introduction to Data Science with Python

The amount of digital data that exists is growing at a rapid rate, doubling every two years, and changing the way we live. It is estimated that by 2020, about 1.7MB of new data will be created every second for every human being on the planet. This means we need to have the technical tools, algorithms, and models to clean, process, and understand the available data in its different forms for decision-making purposes. *Data science* is the field that comprises everything related to cleaning, preparing, and analyzing unstructured, semistructured, and structured data. This field of science uses a combination of statistics, mathematics, programming, problem-solving, and data capture to extract insights and information from data.

The Stages of Data Science

Figure 1-1 shows different stages in the field of data science. Data scientists use programming tools such as Python, R, SAS, Java, Perl, and C/C++ to extract knowledge from prepared data. To extract this information, they employ various fit-to-purpose models based on machine leaning algorithms, statistics, and mathematical methods.

O. Embarak, *Data Analysis and Visualization Using Python*,
https://doi.org/10.1007/978-1-4842-4109-7_1

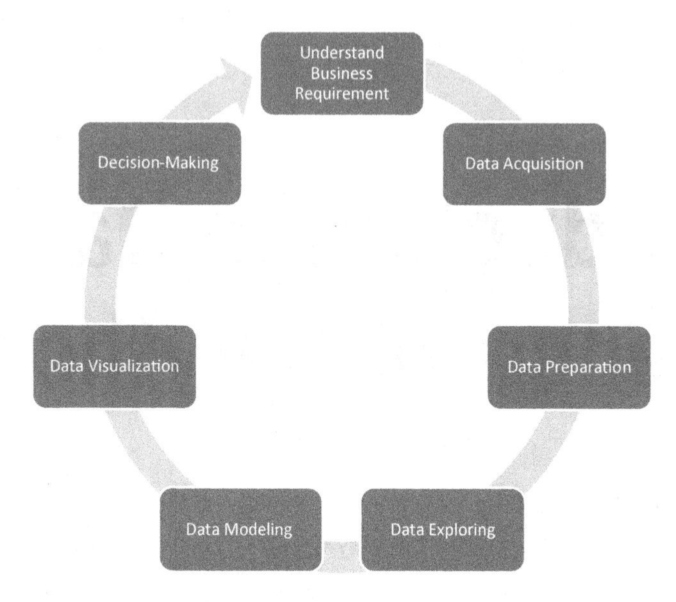

Figure 1-1. *Data science project stages*

Data science algorithms are used in products such as internet search engines to deliver the best results for search queries in less time, in recommendation systems that use a user's experience to generate recommendations, in digital advertisements, in education systems, in healthcare systems, and so on. Data scientists should have in-depth knowledge of programming tools such as Python, R, SAS, Hadoop platforms, and SQL databases; good knowledge of semistructured formats such as JSON, XML, HTML. In addition to the knowledge of how to work with unstructured data.

Why Python?

Python is a dynamic and general-purpose programming language that is used in various fields. Python is used for everything from throwaway scripts to large, scalable web servers that provide uninterrupted service 24/7. It is used for GUI and database programming, client- and server-side web programming, and application testing. It is used by scientists writing applications for the world's fastest supercomputers and by children first learning to program. It was initially developed in the early 1990s by Guido van Rossum and is now controlled by the not-for-profit Python Software Foundation, sponsored by Microsoft, Google, and others.

The first-ever version of Python was introduced in 1991. Python is now at version 3.*x*, which was released in February 2011 after a long period of testing. Many of its major features have also been backported to the backward-compatible Python 2.6, 2.7, and 3.6.

Basic Features of Python

Python provides numerous features; the following are some of these important features:

- *Easy to learn and use*: Python uses an elegant syntax, making the programs easy to read. It is developer-friendly and is a high-level programming language.
- *Expressive*: The Python language is expressive, which means it is more understandable and readable than other languages.
- *Interpreted*: Python is an interpreted language. In other words, the interpreter executes the code line by line. This makes debugging easy and thus suitable for beginners.
- *Cross-platform*: Python can run equally well on different platforms such as Windows, Linux, Unix, Macintosh, and so on. So, Python is a portable language.
- *Free and open source*: The Python language is freely available at `www.python.org`. The source code is also available.

- *Object-oriented*: Python is an object-oriented language with concepts of classes and objects.
- *Extensible*: It is easily extended by adding new modules implemented in a compiled language such as C or C++, which can be used to compile the code.
- *Large standard library*: It comes with a large standard library that supports many common programming tasks such as connecting to web servers, searching text with regular expressions, and reading and modifying files.
- *GUI programming support*: Graphical user interfaces can be developed using Python.
- *Integrated*: It can be easily integrated with languages such as C, C++, Java, and more.

Python Learning Resources

Numerous amazing Python resources are available to train Python learners at different learning levels. There are so many resources out there, though it can be difficult to know how to find all of them. The following are the best general Python resources with descriptions of what they provide to learners:

- *Python Practice Book* is a book of Python exercises to help you learn the basic language syntax. (See `https://anandology.com/python-practice-book/index.html`.)
- *Agile Python Programming: Applied for Everyone* provides a practical demonstration of Python programming as an agile tool for data cleaning, integration, analysis, and visualization fits for academics, professionals, and

researchers. (See `http://www.lulu.com/shop/ossama-embarak/agile-python-programming-applied-for-everyone/paperback/product-23694020.html`.)

- "A Python Crash Course" gives an awesome overview of the history of Python, what drives the programming community, and example code. You will likely need to read this in combination with other resources to really let the syntax sink in, but it's a great resource to read several times over as you continue to learn. (See `https://www.grahamwheeler.com/posts/python-crash-course.html`.)
- "A Byte of Python" is a beginner's tutorial for the Python language. (See `https://python.swaroopch.com/`.)
- The O'Reilly book *Think Python*: *How to Think Like a Computer Scientist* is available in HTML form for free on the Web. (See `https://greenteapress.com/wp/think-python/`.)
- *Python for You and Me* is an approachable book with sections for Python syntax and the major language constructs. The book also contains a short guide at the end teaching programmers to write their first Flask web application. (See `https://pymbook.readthedocs.io/en/latest/`.)
- Code Academy has a Python track for people completely new to programming. (See `www.codecademy.com/catalog/language/python`.)
- *Introduction to Programming with Python* goes over the basic syntax and control structures in Python. The free book has numerous code examples to go along with each topic. (See `www.opentechschool.org/`.)

- Google has a great compilation of material you should read and learn from if you want to be a professional programmer. These resources are useful not only for Python beginners but for any developer who wants to have a strong professional career in software. (See `techdevguide.withgoogle.com`.)
- Looking for ideas about what projects to use to learn to code? Check out the five programming projects for Python beginners at `knightlab.northwestern.edu`.
- There's a Udacity course by one of the creators of Reddit that shows how to use Python to build a blog. It's a great introduction to web development concepts. (See `mena.udacity.com`.)

Python Environment and Editors

Numerous integrated development environments (IDEs) can be used for creating Python scripts.

Portable Python Editors (No Installation Required)

These editors require no installation:

Azure Jupyter Notebooks: The open source Jupyter Notebooks was developed by Microsoft as an analytic playground for analytics and machine learning.

Python(x,y): Python(x,y) is a free scientific and engineering development application for numerical computations, data analysis, and data visualization based on the Python programming language, Qt graphical user interfaces, and Spyder interactive scientific development environment.

WinPython: This is a free Python distribution for the Windows platform; it includes prebuilt packages for ScientificPython.

Anaconda: This is a completely free enterprise-ready Python distribution for large-scale data processing, predictive analytics, and scientific computing.

PythonAnywhere: PythonAnywhere makes it easy to create and run Python programs in the cloud. You can write your programs in a web-based editor or just run a console session from any modern web browser.

Anaconda Navigator: This is a desktop graphical user interface (GUI) included in the Anaconda distribution that allows you to launch applications and easily manage Anaconda packages (as shown in Figure 1-2), environments, and channels without using command-line commands. Navigator can search for packages on the Anaconda cloud or in a local Anaconda repository. It is available for Windows, macOS, and Linux.

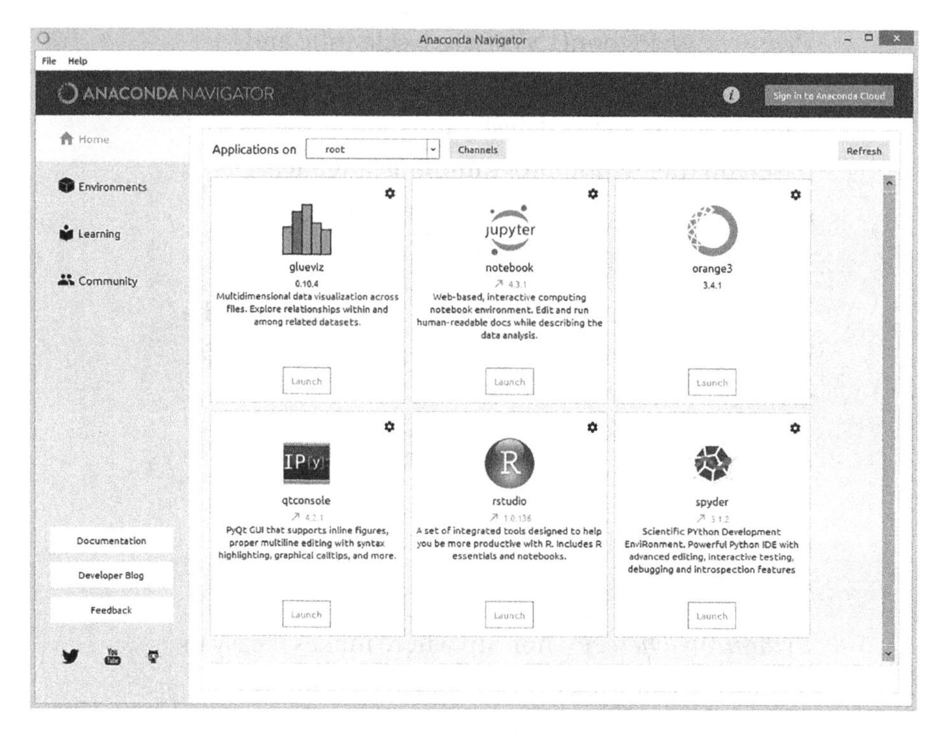

Figure 1-2. *Anaconda Navigator*

The following sections demonstrate how to set up and use Azure Jupyter Notebooks.

Azure Notebooks

The Azure Machine Learning workbench supports interactive data science experimentation through its integration with Jupyter Notebooks.

Azure Notebooks is available for free at `https://notebooks.azure.com/`. After registering and logging into Azure Notebooks, you will get a menu that looks like this:

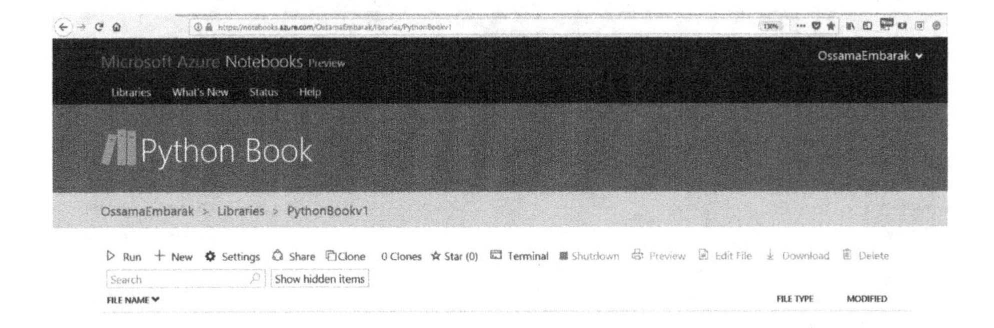

Once you have created your account, you can create a library for any Python project you would like to start. All libraries you create can be displayed and accessed by clicking the Libraries link.

Let's create a new Python script.

1. Create a library.

   Click New Library, enter your library details, and click Create, as shown here:

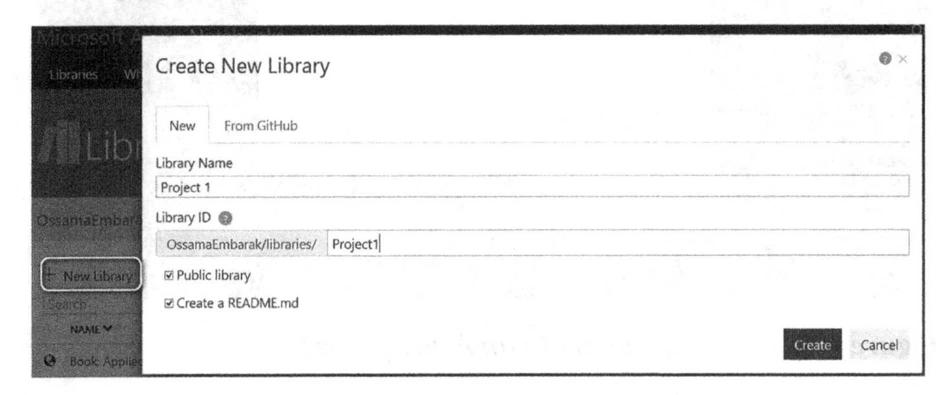

A new library is created, as shown in Figure 1-3.

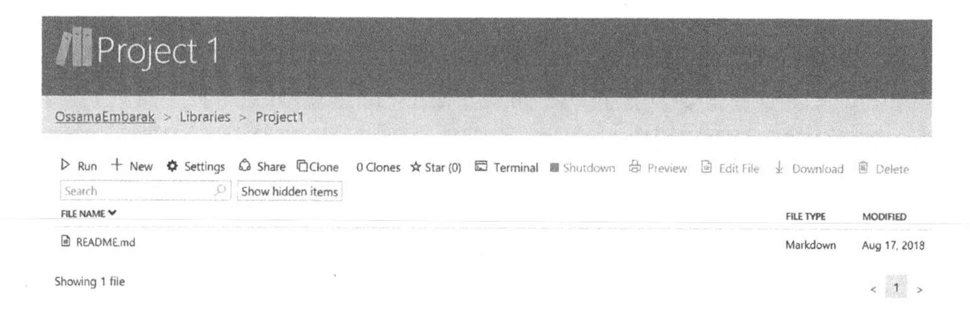

2. Create a project folder container.

   Organizing the Python library scripts is important. You can create folders and subfolders by selecting +New from the ribbon; then for the item type select Folder, as shown in Figure 1-3.

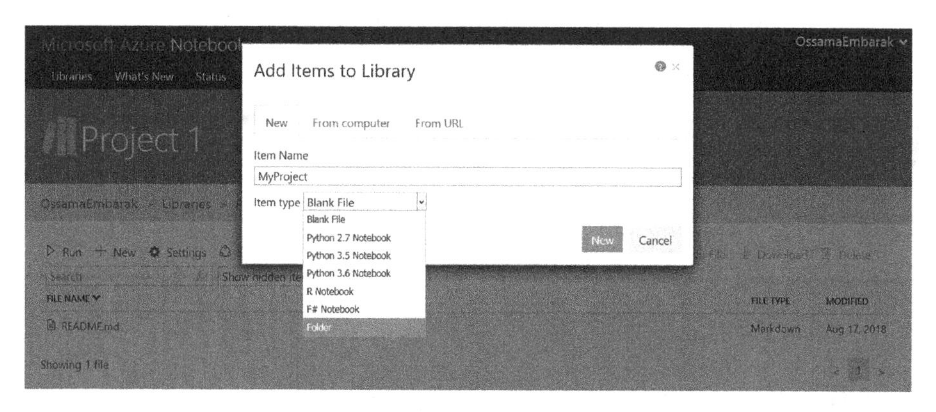

Figure 1-3. *Creating a folder in an Azure project*

3. Create a Python project.

Move inside the created folder and create a new Python project.

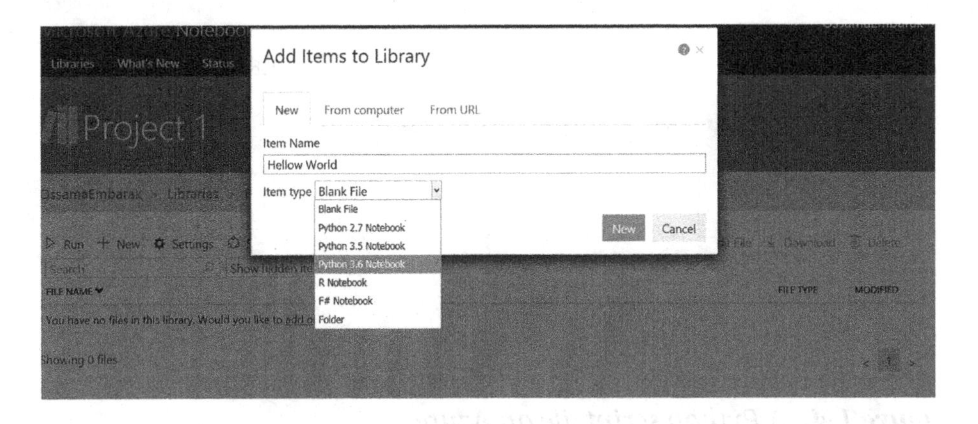

Your project should look like this:

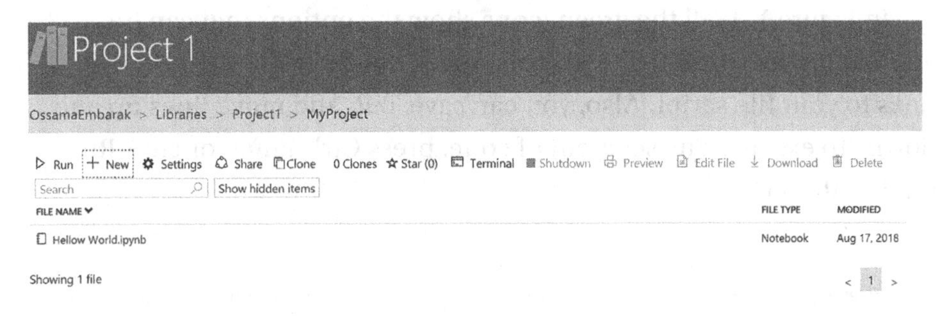

4. Write and run a Python script.

Open the Created Hello World script by clicking it, and start writing your Python code, as shown in Figure 1-4.

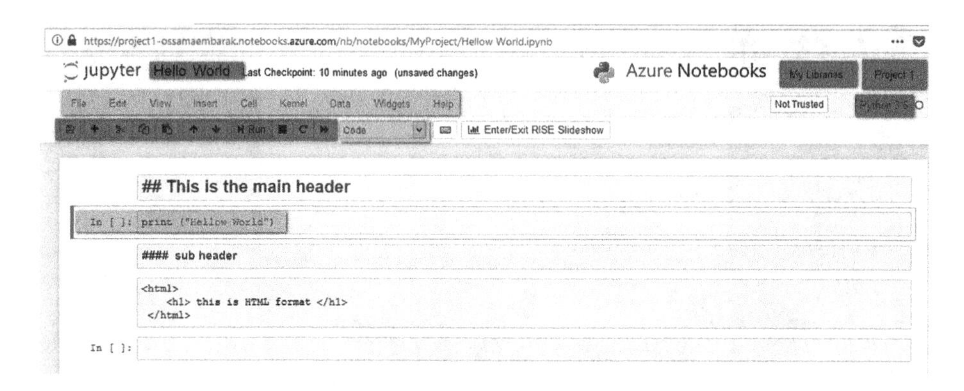

Figure 1-4. *A Python script file on Azure*

In Figure 1-4, all the green icons show the options that can be applied on the running file. For instance, you can click + to add new lines to your file script. Also, you can save, cut, and move lines up and down. To execute any segment of code, press Ctrl+Enter, or click Run on the ribbon.

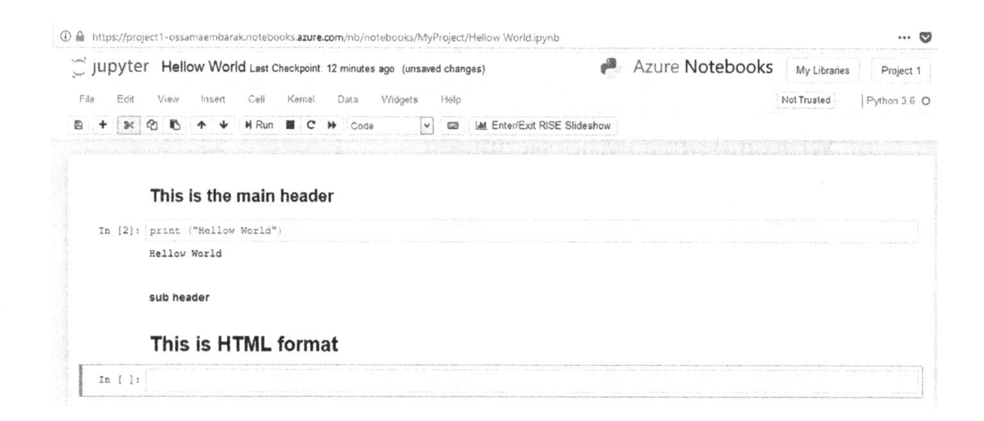

Offline and Desktop Python Editors

There are many offline Python IDEs such as Spyder, PyDev via Eclipse, NetBeans, Eric, PyCharm, Wing, Komodo, Python Tools for Visual Studio, and many more.

The following steps demonstrate how to set up and use Spyder. You can download Anaconda Navigator and then run the Spyder software, as shown in Figure 1-5.

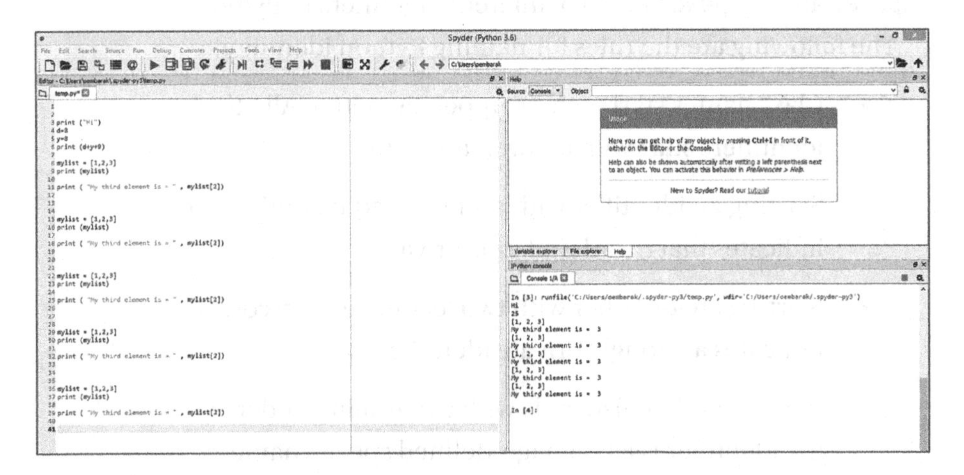

Figure 1-5. *Python Spyder IDE*

On the left side, you can write Python scripts, and on the right side you can see the executed script in the console.

The Basics of Python Programming

This section covers basic Python programming.

Basic Syntax

A Python *identifier* is a name used to identify a variable, function, class, module, or other object in the created script. An identifier starts with a letter from *A* to *Z* or from *a* to *z* or an underscore (_) followed by zero or more letters, underscores, and digits (0 to 9).

Python does not allow special characters such as @, $, and % within identifiers. Python is a case-sensitive programming language. Thus, `Manpower` and `manpower` are two different identifiers in Python.

The following are the rules for naming Python identifiers:

- Class names start with an uppercase letter. All other identifiers start with a lowercase letter.
- Starting an identifier with a single leading underscore indicates that the identifier is private.
- Starting an identifier with two leading underscores indicates a strongly private identifier.
- If the identifier also ends with two trailing underscores, the identifier is a language-defined special name.

The `help?` method can be used to get support from the Python user manual, as shown in Listing 1-1.

Listing 1-1. Getting Help from Python

```
In [3]:      help?

Signature:   help(*args, **kwds)
Type:        _Helper
String form: Type help() for interactive help, or help(object)
for help about object.
Namespace:   Python builtin
```

```
File:         ~/anaconda3_501/lib/python3.6/_sitebuiltins.py
Docstring:
Define the builtin 'help'.

This is a wrapper around pydoc.help that provides a helpful
message
when 'help' is typed at the Python interactive prompt.

Calling help() at the Python prompt starts an interactive help
session.
Calling help(thing) prints help for the python object 'thing'.
```

The smallest unit inside a given Python script is known as a *token*, which represents punctuation marks, reserved words, and each individual word in a statement, which could be keywords, identifiers, literals, and operators.

Table 1-1 lists the reserved words in Python. Reserved words are the words that are reserved by the Python language already and can't be redefined or declared by the user.

Table 1-1. *Python Reserved Keywords*

and	exec	not	continue	global	with	yield	in
assert	finally	or	def	if	return	else	is
break	for	pass	except	lambda	while	try	
class	from	print	del	import	raise	elif	

Lines and Indentation

Line indentation is important in Python because Python does not depend on braces to indicate blocks of code for class and function definitions or flow control. Therefore, a code segment block is denoted by line indentation, which is rigidly enforced, as shown in Listing 1-2.

Listing 1-2. Line Indentation Syntax Error

```
In [4]:age, mark, code=10,75,"CIS2403"
      print (age)
      print (mark)
           print (code)

File "<ipython-input-4-5e544bb51da0>", line 4
print (code)
IndentationError: unexpected indent
```

Multiline Statements

Statements in Python typically end with a new line. But a programmer can use the line continuation character (\) to denote that the line should continue, as shown in Listing 1-3. Otherwise, a syntax error will occur.

Listing 1-3. Multiline Statements

```
In [5]:TV=15
            Mobile=20 Tablet = 30
total = TV +
Mobile +
      Tablet
print (total)

File "<ipython-input-5-68bc7095f603>", line 5
total = TV +
SyntaxError: invalid syntax
```

The following is the correct syntax:

```
In [6]: TV=15
       Mobile=20
       Tablet = 30
       total = TV + \
```

```
        Mobile + \
        Tablet
        print (total)
```

```
65
```

The code segment with statements contained within the [], {}, or () brackets does not need to use the line continuation character, as shown in Listing 1-4.

Listing 1-4. Statements with Quotations

```
In [7]: days = ['Monday', 'Tuesday', 'Wednesday',
'Thursday', 'Friday']
print (days)
```

```
['Monday', 'Tuesday', 'Wednesday', 'Thursday', 'Friday']
```

Quotation Marks in Python

Python accepts single ('), double ("), and triple (''' or """) quotes to denote string literals, as long as the same type of quote starts and ends the string. However, triple quotes are used to span the string across multiple lines, as shown in Listing 1-5.

Listing 1-5. Quotation Marks in Python

```
In [8]:sms1 = 'Hellow World'
      sms2 = "Hellow World"
      sms3 = """ Hellow World"""
      sms4 = """ Hellow
            World"""
      print (sms1)
      print (sms2)
      print (sms3)
      print (sms4)
```

```
Hellow World
Hellow World
Hellow World
Hellow
World
```

Multiple Statements on a Single Line

Python allows the use of \n to split line into multiple lines. In addition, the semicolon (;) allows multiple statements on a single line if neither statement starts a new code block, as shown in Listing 1-6.

Listing 1-6. The Use of the Semicolon and New Line Delimiter

```
In [9]: TV=15; name="Nour"; print (name); print ("Welcome
to\nDubai Festival 2018")

Nour
Welcome to
Dubai Festival 2018
```

Read Data from Users

The line code segment in Listing 1-7 prompts the user to enter a name and age, converts the age into an integer, and then displays the data.

Listing 1-7. Reading Data from the User

```
In [10]:name = input("Enter your name ")
      age = int (input("Enter your age "))
      print ("\nName =", name); print ("\nAge =", age)
```

```
Enter your name Nour
Enter your age 12
Name = Nour
Age = 12
```

Declaring Variables and Assigning Values

There is no restriction to declaring explicit variables in Python. Once you assign a value to a variable, Python considers the variable according to the assigned value. If the assigned value is a string, then the variable is considered a string. If the assigned value is a real, then Python considers the variable as a double variable. Therefore, Python does not restrict you to declaring variables before using them in the application. It allows you to create variables at the required time.

Python has five standard data types that are used to define the operations possible on them and the storage method for each of them.

- Number
- String
- List
- Tuple
- Dictionary

The equal (=) operator is used to assign a value to a variable, as shown in Listing 1-8.

Listing 1-8. Assign Operator

```
In [11]: age = 11
      name ="Nour"
      tall=100.50
In [12]: print (age)
      print (name)
      print (tall)

11
Nour
100.5
```

Multiple Assigns

Python allows you to assign a value to multiple variables in a single statement, which is also known as *multiple assigns*. You can assign a single value to multiple variables or assign multiple values to multiple variables, as shown in Listing 1-9.

Listing 1-9. Multiple Assigns

```
In [13]:age= mark = code =25
      print (age)
      print (mark)
      print (code)

25
25
25

In [14]:age, mark, code=10,75,"CIS2403"
      print (age)
      print (mark)
      print (code)
```

```
10
75
CIS2403
```

Variable Names and Keywords

A *variable* is an identifier that allocates specific memory space and assigns a value that could change during the program runtime. Variable names should refer to the usage of the variable, so if you want to create a variable for student age, then you can name it as age or `student_age`. There are many rules and restrictions for variable names. It's not allowed to use special characters or white spaces in variable naming. For instance, variable names shouldn't start with any special character and shouldn't be any of the Python reserved keywords. The following example shows incorrect naming: `{?age, 1age, age student, and, if, 1_age, etc}`. The following shows correct naming for a variable: `{age, age1, age_1, if_age, etc}`.

Statements and Expressions

A *statement* is any unit of code that can be executed by a Python interpreter to get a specific result or perform a specific task. A program contains a sequence of statements, each of which has a specific purpose during program execution. The expression is a combination of values, variables, and operators that are evaluated by the interpreter to do a specific task, as shown in Listing 1-10.

Listing 1-10. Expression and Statement Forms

```
In [16]:# Expressions
      x=0.6                 # Statement
      x=3.9 * x * (1-x)  # Expressions
      print (round(x, 2))

0.94
```

Basic Operators in Python

Operators are the constructs that can manipulate the value of operands. Like different programming languages, Python supports the following operators:

- Arithmetic operators
- Relational operators
- Assign operators
- Logical operators
- Membership operators
- Identity operators
- Bitwise operators

Arithmetic Operators

Table 1-2 shows examples of arithmetic operators in Python.

Table 1-2. *Python Arithmetic Operators*

Operators	Description	Example	Output
//	Performs floor division (gives the integer value after division)	`print (13//5)`	2
+	Performs addition	`print (13+5)`	18
-	Performs subtraction	`print (13-5)`	8
*	Performs multiplication	`print (2*5)`	10
/	Performs division	`print (13/5)`	2.6
%	Returns the remainder after division (modulus)	`print (13%5)`	3
**	Returns an exponent (raises to a power)	`print (2**3)`	8

Relational Operators

Table 1-3 shows examples of relational operators in Python.

Table 1-3. *Python Relational Operators*

Operators	Description	Example	Output
<	Less than	`print (13<5)`	False
>	Greater than	`print (13>5)`	True
<=	Less than or equal to	`print (13<=5)`	False
>=	Greater than or equal to	`print (2>=5)`	False
==	Equal to	`print (13==5)`	False
!=	Not equal to	`print (13! =5)`	True

Assign Operators

Table 1-4 shows examples of assign operators in Python.

Table 1-4. *Python Assign Operators*

Operators	Description	Example	Output
=	Assigns	`x=10` `print (x)`	10
/=	Divides and assigns	`x=10; x/=2` `print (x)`	5.0
+=	Adds and assigns	`x=10; x+=7` `print (x)`	17
-=	Subtracts and assigns	`x=10; x-=6` `print (x)`	4

(continued)

Table 1-4. *(continued)*

Operators	Description	Example	Output
*=	Multiplies and assigns	`x=10; x*=5` `print (x)`	50
%=	Modulus and assigns	`x=13; x%=5` `print (x)`	3
**=	Exponent and assigns	`x=10; x**=3` `print(x)`	1000
//=	Floor division and assigns	`x=10; x//=2` `print(x)`	5

Logical Operators

Table 1-5 shows examples of logical operators in Python.

Table 1-5. *Python Logical Operators*

Operators	Description	Example	Output
and	Logical AND (when both conditions are true, the output will be true)	`x=10>5 and 4>20` `print (x)`	False
or	Logical OR (if any one condition is true, the output will be true)	`x=10>5 or 4>20` `print (x)`	True
not	Logical NOT (complements the condition; i.e., reverses it)	`x=not (10<4)` `print (x)`	True

A Python *program* is a sequence of Python statements that have been crafted to do something. It can be one line of code or thousands of code segments written to perform a specific task by a computer. Python statements are executed immediately and do not wait for the entire program to be executed. Therefore, Python is an interpreted language that executes line per line. This differs from other languages such as C#, which is a compiled language that needs to handle the entire program.

Python Comments

There are two types of comments in Python: single-line comments and multiline comments.

The # symbol is used for single-line comments.

Multiline comments can be given inside triple quotes, as shown in Listing 1-11.

Listing 1-11. Python Comment Forms

```
In [18]: # Python single line comment
In [19]: ''' This
      Is
      Multi-line comment'''
```

Formatting Strings

The Python special operator % helps to create formatted output. This operator takes two operands, which are a formatted string and a value. The following example shows that you pass a string and the 3.14259 value in string format. It should be clear that the value can be a single value, a tuple of values, or a dictionary of values.

```
In [20]: print ("pi=%s"%"3.14159")

pi=3.14159
```

Conversion Types

You can convert values using different conversion specifier syntax, as summarized in Table 1-6.

Table 1-6. *Conversion Syntax*

Syntax	Description
%c	Converts to a single character
%d, %i	Converts to a signed decimal integer or long integer
%u	Converts to an unsigned decimal integer
%e, %E	Converts to a floating point in exponential notation
%f	Converts to a floating point in fixed-decimal notation
%g	Converts to the value shorter of %f and %e
%G	Converts to the value shorter of %f and %E
%o	Converts to an unsigned integer in octal
%r	Converts to a string generated with `repr()`
%s	Converts to a string using the `str()` function
%x, %X	Converts to an unsigned integer in hexadecimal

For example, the conversion specifier %s says to convert the value to a string. Therefore, to print a numerical value inside string output, you can use, for instance, `print("pi=%s" % 3.14159)`. You can use multiple conversions within the same string, for example, to convert into double, float, and so on.

```
In [1]:print("The value of %s is = %02f" % ("pi", 3.14159))

The value of pi is = 3.141590
```

You can use a dot (.) followed by a positive integer to specify the precision. In the following example, you can use a tuple of different data types and inject the output in a string message:

```
In [21]:print ("Your name is %s, and your height is %.2f while
your weight is %.2d" % ('Ossama', 172.156783, 75.56647))
```

```
Your name is Ossama, and your height is 172.16 while your
weight is 75
```

In the previous example, you can see that `%.2f` is replaced with the value 172.16 with two decimal fractions after the decimal point, while `%2d` is used to display decimal values only but in a two-digit format.

You can display values read directly from a dictionary, as shown next, where `%(name)s` says to take as a string the dictionary value of the key `Name` and `%(height).2f` says to take it as a float with two fraction values, which are the dictionary values of the key `height`:

```
In [23]:print ("Hi %(Name)s, your height is %(height).2f"
%{'Name':"Ossama", 'height': 172.156783})
```

```
Hi Ossama, your height is 172.16
```

The Replacement Field, {}

You can use the replacement field, `{}`, as a name (or index). If an index is provided, it is the index of the list of arguments provided in the field. It's not necessary to have indices with the same sequence; they can be in a random order, such as indices 0, 1, and 2 or indices 2, 1, and 0.

```
In [24]:x = "price is"
        print ("{1} {0} {2}".format(x, "The", 1920.345))
```

```
The price is 1920.345
```

Also, you can use a mix of values combined from lists, dictionaries, attributes, or even a singleton variable. In the following example, you will create a class called `A()`, which has a single variable called `x` that is assigned the value 9.

Then you create an instance (*object*) called `w` from the class `A()`. Then you print values indexed from variable `{0}` and the `{1[2]}` value from the list of values `["a," "or," "is"]`, where 1 refers to the index of printing and 2 refers to the index in the given list where the string index is 0. `{2[test]}` refers to index 2 in the print string and reads its value from the passed dictionary from the key `test`. Finally, `{3.x}` refers to the third index, which takes its value from `w`, which is an instance of the class `A()`.

```
In [34]:class A():x=9 w=A()
        print ("{0} {1[2]} {2[test]} {3.x}".format("This", ["a",
        "or", "is"], {"test": "another"},w))
```

```
This is another 9
```

```
In [34]:print ("{1[1]} {0} {1[2]} {2[test]}{3.x}".
format("This", ["a", "or", "is"], {"test": "another"},w))
```

```
or This is another 9
```

The Date and Time Module

Python provides a time package to deal with dates and times. You can retrieve the current date and time and manipulate the date and time using the built-in methods.

The example in Listing 1-12 imports the time package and calls its `.localtime()` function to retrieve the current date and time.

Listing 1-12. Time Methods

```
In [42]:import time localtime = time.asctime(time.
localtime(time.time()))
print ("Formatted time :", localtime)
print(time.localtime())
print (time.time())

Formatted time : Fri Aug 17 19:12:07 2018

time.struct_time(tm_year=2018, tm_mon=8, tm_mday=17,
tm_hour=19, tm_min=12, tm_sec=7, tm_wday=4, tm_yday=229,
tm_isdst=0)

1534533127.8304486
```

Time Module Methods

Python provides various built-in time functions, as in Table 1-7, that can be used for time-related purposes.

Table 1-7. *Built-in Time Methods*

Methods	Description
time()	Returns time in seconds since January 1, 1970.
asctime(time)	Returns a 24-character string, e.g., Sat Jun 16 21:27:18 2018.
sleep(time)	Used to stop time for the given interval of time.
strptime (String,format)	Returns a tuple with nine time attributes. It receives a string of date and a format. `time.struct_time(tm_year=2018, tm_mon=6, tm_mday=16, tm_hour=0, tm_min=0, tm_sec=0, tm_wday=3, tm_yday=177, tm_isdst=-1)`

(continued)

Table 1-7. *(continued)*

Methods	Description
gtime()/ gtime(sec)	Returns struct_time, which contains nine time attributes.
mktime()	Returns the seconds in floating point since the epoch.
strftime (format)/ strftime (format,time)	Returns the time in a particular format. If the time is not given, the current time in seconds is fetched.

Python Calendar Module

Python provides a calendar module, as in Table 1-8, which provides many functions and methods to work with a calendar.

Table 1-8. *Built-in Calendar Module Functions*

Methods	Description
prcal(year)	Prints the whole calendar of the year.
firstweekday()	Returns the first weekday. It is by default 0, which specifies Monday.
isleap(year)	Returns a Boolean value, i.e., true or false. Returns true in the case the given year is a leap year; otherwise, false.
monthcalendar(year,month)	Returns the given month with each week as one list.
leapdays(year1,year2)	Returns the number of leap days from year1 to year2.
prmonth(year,month)	Prints the given month of the given year.

You can use the Calendar package to display a 2018 calendar as shown here:

```
In [45]:import calendar
         calendar.prcal(2018)
```

```
                                  2018

      January                   February                   March
Mo Tu We Th Fr Sa Su      Mo Tu We Th Fr Sa Su      Mo Tu We Th Fr Sa Su
 1  2  3  4  5  6  7                1  2  3  4                1  2  3  4
 8  9 10 11 12 13 14       5  6  7  8  9 10 11       5  6  7  8  9 10 11
15 16 17 18 19 20 21      12 13 14 15 16 17 18      12 13 14 15 16 17 18
22 23 24 25 26 27 28      19 20 21 22 23 24 25      19 20 21 22 23 24 25
29 30 31                  26 27 28                  26 27 28 29 30 31

       April                      May                       June
Mo Tu We Th Fr Sa Su      Mo Tu We Th Fr Sa Su      Mo Tu We Th Fr Sa Su
                   1          1  2  3  4  5  6                   1  2  3
 2  3  4  5  6  7  8       7  8  9 10 11 12 13       4  5  6  7  8  9 10
 9 10 11 12 13 14 15      14 15 16 17 18 19 20      11 12 13 14 15 16 17
16 17 18 19 20 21 22      21 22 23 24 25 26 27      18 19 20 21 22 23 24
23 24 25 26 27 28 29      28 29 30 31               25 26 27 28 29 30
30

        July                     August                  September
Mo Tu We Th Fr Sa Su      Mo Tu We Th Fr Sa Su      Mo Tu We Th Fr Sa Su
                   1             1  2  3  4  5                      1  2
 2  3  4  5  6  7  8       6  7  8  9 10 11 12       3  4  5  6  7  8  9
 9 10 11 12 13 14 15      13 14 15 16 17 18 19      10 11 12 13 14 15 16
16 17 18 19 20 21 22      20 21 22 23 24 25 26      17 18 19 20 21 22 23
23 24 25 26 27 28 29      27 28 29 30 31            24 25 26 27 28 29 30
30 31

      October                   November                  December
Mo Tu We Th Fr Sa Su      Mo Tu We Th Fr Sa Su      Mo Tu We Th Fr Sa Su
 1  2  3  4  5  6  7                1  2  3  4                      1  2
 8  9 10 11 12 13 14       5  6  7  8  9 10 11       3  4  5  6  7  8  9
15 16 17 18 19 20 21      12 13 14 15 16 17 18      10 11 12 13 14 15 16
22 23 24 25 26 27 28      19 20 21 22 23 24 25      17 18 19 20 21 22 23
29 30 31                  26 27 28 29 30            24 25 26 27 28 29 30
                                                    31
```

Fundamental Python Programming Techniques

This section demonstrates numerous Python programming syntax structures.

Selection Statements

The `if` statement is used to execute a specific statement or set of statements when the given condition is true. There are various forms of `if` structures, as shown in Table 1-9.

Table 1-9. *if Statement Structure*

Form	if statement	if-else Statement	Nested if Statement

Structure	`if(condition):`
	`statements`
		`if(condition):`
		`statements`
		`else:`
		`statements`
			`if (condition):`
			`statements`
			`elif (condition):`
			`statements`
			`else:`
			`statements`

The `if` statement is used to make decisions based on specific conditions occurring during the execution of the program. An action or set of actions is executed if the outcome is true or false otherwise. Figure 1-6 shows the general form of a typical decision-making structure found in most programming languages including Python. Any nonzero and non-null values are considered true in Python, while either zero or null values are considered false.

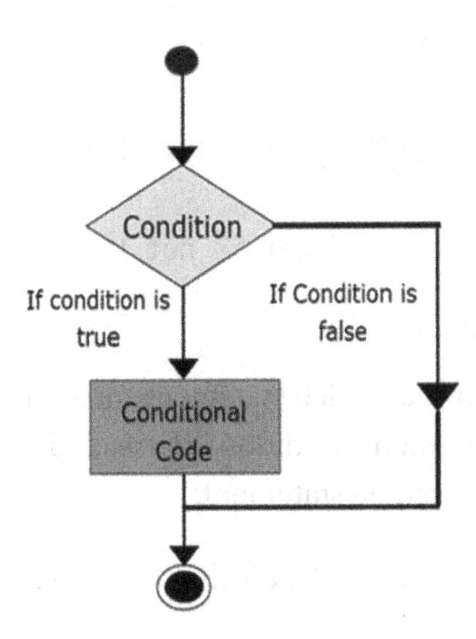

Figure 1-6. *Selection statement structure*

Listing 1-13 demonstrates two examples of a selection statement, remember the indentation is important in the Python structure. The first block shows that the value of x is equal to 5; hence, the condition is testing whether x equals 5 or not. Therefore, the output implements the statement when the condition is true.

Listing 1-13. The if-else Statement Structure

```
In [13]:#Comparison operators
      x=5
      if x==5:
             print ('Equal 5')
elif x>5:
      print ('Greater than 5')
elif x<5:
      print ('Less than 5')

Equal 5
```

```
In [14]:year=2000
        if year%4==0:
                print("Year(", year ,")is Leap")
else:
                print (year , "Year is not Leap" )
```

```
Year( 2000 )is Leap
```

Indentation determines which statement should be executed. In Listing 1-14, the `if` statement condition is false, and hence the outer `print` statement is the only executed statement.

Listing 1-14. Indentation of Execution

```
In [12]:#Indentation
        x=2
        if x>2:
                 print ("Bigger than 2")
                 print (" X Value bigger than 2")
        print ("Now we are out of if block\n")
```

```
Now we are out of if block
```

The nested `if` statement is an `if` statement that is the target of another `if` statement. In other words, a nested `if` statement is an `if` statement inside another `if` statement, as shown in Listing 1-15.

Listing 1-15. Nested Selection Statements

```
In [2]:a=10
        if a>=20:
                 print ("Condition is True" )
else:
                 if a>=15:
                         print ("Checking second value" )
```

```
else:
                  print ("All Conditions are false" )
```

```
All Conditions are false
```

Iteration Statements

There are various iteration statement structures in Python. The `for` loop is one of these structures; it is used to iterate the elements of a collection in the order that they appear. In general, statements are executed sequentially, where the first statement in a function is executed first, followed by the second, and so on. There may be a situation when you need to execute a block of code several numbers of times.

Control structures allow you to execute a statement or group of statements multiple times, as shown by Figure 1-7.

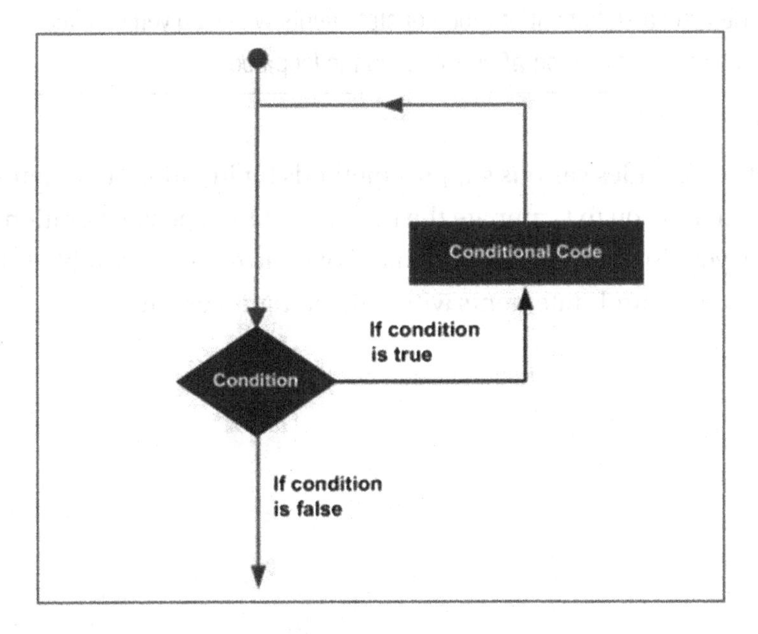

Figure 1-7. *A loop statement*

Table 1-10 demonstrates different forms of iteration statements. The Python programming language provides different types of loop statements to handle iteration requirements.

Table 1-10. *Iteration Statement Structure*

1	`for` **loop** Executes a sequence of statements multiple times and abbreviates the code that manages the loop variable.
2	**Nested loops** You can use one or more loop inside any another `while`, `for`, or `do..while` loop.
3	`while` **loop** Repeats a statement or group of statements while a given condition is true. It tests the condition *before* executing the loop body.
4	`do {....} while ()` Repeats a statement or group of statements while a given condition is true. It tests the condition *after* executing the loop body.

Python provides various support methods for iteration statements where it allows you to terminate the iteration, skip a specific iteration, or pass if you do not want any command or code to execute. Table 1-11 summarizes control statements within the iteration execution.

Table 1-11. Loop Control Statements

1	**Break statement** Terminates the loop statement and transfers execution to the statement immediately following the loop.
2	**Continue statement** Causes the loop to skip the remainder of its body and immediately retests its condition prior to reiterating.
3	**Pass statement** The pass statement is used when a statement is required syntactically but you do not want any command or code to execute.

The `range()` statement is used with `for` loop statements where you can specify one value. For example, if you specify 4, the loop statement starts from 1 and ends with 3, which is n-1. Also, you can specify the start and end values. The following examples demonstrate loop statements.

Listing 1-16 displays all numerical values starting from 1 up to n-1, where n=4.

Listing 1-16. for Loop Statement

```
In [23]:# use the range statement
       for a in range (1,4):
       print ( a )

1
2
3
```

Listing 1-17 displays all numerical values starting from 0 up to n-1, where n=4.

Listing 1-17. Using the range() Method

```
In [24]:# use the range statement
      for a in range (4):
            print ( a )

0
1
2
3
```

Listing 1-18 displays the `while` iteration statement.

Listing 1-18. while Iteration Statement

```
In [32]:ticket=4
      while ticket>0:
            print ("Your ticket number is ", ticket)
            ticket -=1

Your ticket number is 4
Your ticket number is 3
Your ticket number is 2
Your ticket number is 1
```

Listing 1-19 iterates all numerical values in a list to find the maximum value.

Listing 1-19. Using a Selection Statement Inside a Loop Statement

```
In [2]:largest = None
      print ('Before:', largest)
      for val in [30, 45, 12, 90, 74, 15]:
if largest is None or val>largest:
      largest = val
      print ("Loop", val, largest)
print ("Largest", largest)
```

```
Before: None
Loop 30 30
Loop 45 45
Loop 90 90
Largest 90
```

In the previous examples, the first and second iterations used the for loop with a range statement. In the last example, iteration goes through a list of elements and stops once it reaches the last element of the iterated list.

A *break* statement is used to jump statements and transfer the execution control. It breaks the current execution, and in the case of an inner loop, the inner loop terminates immediately. However, a *continue* statement is a jump statement that skips execution of current iteration. After skipping, the loop continues with the next iteration. The pass keyword is used to execute nothing. The following examples demonstrate how and when to employ each statement.

The Use of Break, Continues, and Pass Statements

Listing 1-20 shows the break, continue, and pass statements.

Listing 1-20. Break, Continue, and Pass Statements

```
In [44]:for letter in 'Python3':
       if letter == 'o':
             break
        print (letter)
```

```
P
y
t
h

In [45]: a=0
        while a<=5:
             a=a+1
             if a%2==0:
              continue
             print (a)
 print ("End of Loop" )

1
3
5
End of Loop

In [46]: for i in [1,2,3,4,5]:
              if i==3:
                   pass
              print ("Pass when value is", i )
            print (i)

1
2
Pass when value is 3
3
4
5
```

As shown, you can iterate over a list of letters, as shown in Listing 1-20, and you can iterate over the word *Python3* and display all the letters. You stop iteration once you find the condition, which is the letter *o*. In addition, you can use the pass statement when a statement is required syntactically but you do not want any command or code to execute. The pass statement is a null operation; nothing happens when it executes.

try and except

`try` and `except` are used to handle unexpected values where you would like to validate entered values to avoid error occurrence. In the first example of Listing 1-21, you use `try` and `except` to handle the string "Al Fayoum," which is not convertible into an integer, while in the second example, you use `try` and `except` to handle the string 12, which is convertible to an integer value.

Listing 1-21. try and except Statements

```
In [14]: # Try and Except
astr='Al Fayoum'
      errosms=''
try:
             istr=int(astr) # error
except:
             istr=-1
             errosms="\nIncorrect entry"
print ("First Try:", istr , errosms)

First Try: -1
Incorrect entry

In [15]:# Try and Except
                   astr='12'
                   errosms=' '
                   try:
                    istr=int(astr) # error
                   except:
```

```
istr=-1
errosms="\nIncorrect entry"
                    print ("First Try:", istr , errosms)
```

```
First Try: 12
```

String Processing

A *string* is a sequence of characters that can be accessed by an expression in brackets called an *index*. For instance, if you have a string variable named `var1`, which maintains the word PYTHON, then `var1[1]` will return the character *Y*, while `var1[-2]` will return the character *O*. Python considers strings by enclosing text in single as well as double quotes. Strings are stored in a contiguous memory location that can be accessed from both directions (forward and backward), as shown in the following example, where

- Forward indexing starts with 0, 1, 2, 3, and so on.
- Backward indexing starts with -1, -2, -3, -4, and so on.

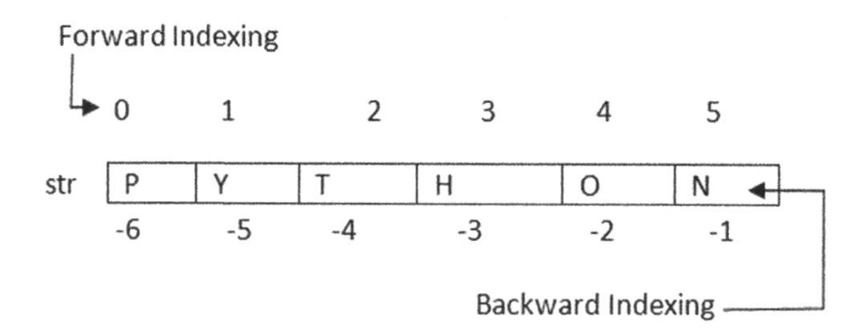

String Special Operators

Table 1-12 lists the operators used in string processing. Say you have the two variables `a= 'Hello'` and `b = 'Python'`. Then you can implement the operations shown in Table 1-12.

Table 1-12. *String Operators*

Operator	Description	Outputs
+	Concatenation: adds values on either side of the operator	a + b will give HelloPython.
*	Repetition: creates new strings, concatenating multiple copies of the same string	a*2 will give -HelloHello.
[]	Slice: gives the character from the given index	a[1] will give e.
[:]	Range slice: gives the characters from the given range	a[1:4] will give ell.
in	Membership: returns true if a character exists in the given string	H in a will give true.
not in	Membership: returns true if a character does not exist in the given string	M not in a will give true.

Various symbols are used for string formatting using the operator %. Table 1-13 gives some simple examples.

Table 1-13. *String Format Symbols*

Format Symbol	Conversion
%c	Character
%s	String conversion via str() prior to formatting
%i	Signed decimal integer
%d	Signed decimal integer
%u	Unsigned decimal integer

(continued)

Table 1-13. *(continued)*

Format Symbol	Conversion
%o	Octal integer
%x	Hexadecimal integer (lowercase letters)
%X	Hexadecimal integer (uppercase letters)
%e	Exponential notation (with lowercase e)
%E	Exponential notation (with uppercase E)
%f	Floating-point real number
%g	The shorter of %f and %e
%G	The shorter of %f and %E

String Slicing and Concatenation

String slicing refers to a segment of a string that is extracted using an index or using search methods. In addition, the `len()` method is a built-in function that returns the number of characters in a string. Concatenation enables you to join more than one string together to form another string.

The operator `[n:m]` returns the part of the string from the *n*th character to the *m*th character, including the first but excluding the last. If you omit the first index (before the colon), the slice starts at the beginning of the string. In addition, if you omit the second index, the slice goes to the end of the string. The examples in Listing 1-22 show string slicing and concatenation using the + operator.

Listing 1-22. String Slicing and Concatenation

```
In [3]:var1 = 'Welcome to Dubai'
      var2 = "Python Programming"
      print ("var1[0]:", var1[0])
      print ("var2[1:5]:", var2[1:5])

      var1[0]: W
      var2[1:5]: ytho

In [5]:st1="Hello"
      st2=' World'
      fullst=st1 + st2
      print (fullst)

Hello World

In [11]:# looking inside strings
      fruit = 'banana'
      letter= fruit[1]
      print (letter)
      index=3
      w = fruit[index-1]
      print (w)
      print (len(fruit))

a
n
6
```

String Conversions and Formatting Symbols

It is possible to convert a string value into a float, double, or integer if the string value is applicable for conversion, as shown in Listing 1-23.

Listing 1-23. String Conversion and Format Symbols

```
In [14]:#Convert string to int
      str3 = '123'
      str3= int (str3)+1
      print (str3)

124

In [15]:#Read and convert data
      name=input('Enter your name: ')
      age=input('Enter your age: ')
      age= int(age) + 1
      print ("Name: %s"% name ,"\t Age:%d"% age)

Enter your name: Omar
Enter your age: 41

Name: Omar    Age:42
```

Loop Through String

You can use iteration statements to go through a string forward or backward. A lot of computations involve processing a string one character at a time. String processing can start at the beginning, select each character in turn, do something to it, and continue until the end. This pattern of processing is called a *traversal*. One way to write a traversal is with a `while` loop, as shown in Listing 1-24.

Listing 1-24. Iterations Through Strings

```
In [30]:# Looking through string
        fruit ='banana'
        index=0
        while index< len(fruit):
               letter = fruit [index]
```

```
            print (index, letter)
            index=index+1
0 b
1 a
2 n
3 a
4 n
5 a

In [31]:print ("\n Implementing iteration with continue")
      while True:
            line = input('Enter your data>')
            if line[0]=='#':
                  continue
            if line =='done':
                  break
            print (line )
      print ('End!')

Implementing iteration with continue

Enter your data>Higher Colleges of Technology
Higher Colleges of Technology

Enter your data>#

Enter your data>done
End!

In [32]:print ("\nPrinting in reverse order")
      index=len(fruit)-1
      while index>=0 :
            letter = fruit [index]
            print (index, letter )
            index=index-1
```

```
Printing in reverse order
5 a
4 n
3 a
2 n
1 a
0 b
```

```
Letterwise iteration
In [33]:Country='Egypt'
      for letter in Country:
            print (letter)

E
g
y
p
t
```

You can use iterations as well to count letters in a word or to count words in lines, as shown in Listing 1-25.

Listing 1-25. Iterating and Slicing a String

```
In [2]:# Looking and counting
      word='banana'
      count=0
      for letter in word:
            if letter =='a':
                  count +=1
      print ("Number of a in ", word, "is :", count )

Number of a in banana is : 3
```

```
In [3]:# String Slicing
      s="Welcome to Higher Colleges of Technology"
      print (s[0:4])
      print (s[6:7])
      print (s[6:20])
      print (s[:12])
      print (s[2:])
      print (s [:])
      print (s)

Welc
e
e to Higher Co Welcome to H
lcome to Higher Colleges of Technology Welcome to Higher
Colleges of Technology
Welcome to Higher Colleges of Technology
```

Python String Functions and Methods

Numerous built-in methods and functions can be used for string processing; Table 1-14 lists these methods.

Table 1-14. *Built-in String Methods*

Method/Function	Description
`capitalize()`	Capitalizes the first character of the string.
`count(string, begin,end)`	Counts a number of times a substring occurs in a string between the beginning and end indices.
`endswith(suffix, begin=0,end=n)`	Returns a Boolean value if the string terminates with a given suffix between the beginning and end.

(continued)

Table 1-14. (*continued*)

Method/Function	Description
`find(substring, beginIndex, endIndex)`	Returns the index value of the string where the substring is found between the begin index and the end index.
`index(subsring, beginIndex, endIndex)`	Throws an exception if the string is not found and works same as the `find()` method.
`isalnum()`	Returns true if the characters in the string are alphanumeric (i.e., letters or numbers) and there is at least one character. Otherwise, returns false.
`isalpha()`	Returns true when all the characters are letters and there is at least one character; otherwise, false.
`isdigit()`	Returns true if all the characters are digits and there is at least one character; otherwise, false.
`islower()`	Returns true if the characters of a string are in lowercase; otherwise, false.
`isupper()`	Returns false if the characters of a string are in uppercase; otherwise, false.
`isspace()`	Returns true if the characters of a string are white space; otherwise, false.
`len(string)`	Returns the length of a string.
`lower()`	Converts all the characters of a string to lowercase.
`upper()`	Converts all the characters of a string to uppercase.
`startswith(str, begin=0,end=n)`	Returns a Boolean value if the string starts with the given `str` between the beginning and end.

(*continued*)

Table 1-14. (*continued*)

Method/Function	Description
swapcase()	Inverts the case of all characters in a string.
lstrip()	Removes all leading white space of a string and can also be used to remove a particular character from leading white spaces.
rstrip()	Removes all trailing white space of a string and can also be used to remove a particular character from trailing white spaces.

Listing 1-26 shows how to use built-in methods to remove white space from a string, count specific letters within a string, check whether the string contains another string, and so on.

Listing 1-26. Implementing String Methods

```
In [29]:var1 =' Higher Colleges of Technology '
      var2='College'
      var3='g'
      print (var1.upper())
      print (var1.lower())
      print ('WELCOME TO'.lower())
      print (len(var1))
      print (var1.count(var3, 2, 29) ) # find how many g
      letters in var1
      print ( var2.count(var3) )

HIGHER COLLEGES OF TECHNOLOGY
higher colleges of technology
welcome to
```

```
31
3
1

In [33]:print (var1.endswith('r'))
      print (var1.startswith('O'))
      print (var1.find('h', 0, 29))
      print (var1.lstrip()) # It removes all leading whitespace
      of a string in var1
      print (var1.rstrip()) # It removes all trailing
      whitespace of a string in var1
      print (var1.strip()) # It removes all leading and
      trailing whitespace
      print ('\n')
      print (var1.replace('Colleges', 'University'))

False
False
4
Higher Colleges of Technology
 Higher Colleges of Technology
Higher Colleges of Technology

Higher University of Technology
```

The in Operator

The word in is a Boolean operator that takes two strings and returns true if the first appears as a substring in the second, as shown in Listing 1-27.

Listing 1-27. The in Method in String Processing

```
In [43]:var1 =' Higher Colleges of Technology '
      var2='College'
      var3='g'
```

```
        print ( var2 in var1)
        print ( var2 not in var1)

True
False
```

Parsing and Extracting Strings

The `find` operator returns the index of the first occurrence of a substring in another string, as shown in Listing 1-28. The `atpost` variable is used to maintain a returned index of the substring `@` as it appears in the `Maindata` string variable.

Listing 1-28. Parsing and Extracting Strings

```
In [39]:# Parsing and Extracting strings
            Maindata = 'From ossama.embarak@hct.ac.ae Sunday
            Jan 4 09:30:50 2017' atpost = Maindata.find('@')
            print ("\n<<<<<<<<<<<<<<>>>>>>>>>>>>>")
            print (atpost)
            print (Maindata[ :atpost])
            data = Maindata[ :atpost]
            name=data.split(' ')
            print (name)
            print (name[1].replace('.', ' ').upper())
            print ("\n<<<<<<<<<<<<<<>>>>>>>>>>>>>")

<<<<<<<<<<<<<<>>>>>>>>>>>>>
19
From ossama.embarak
['From', 'ossama.embarak']
OSSAMA EMBARAK
<<<<<<<<<<<<<<>>>>>>>>>>>>>
```

```
In [41]:# Another way to split strings
         Maindata = 'From ossama.embarak@hct.ac.ae Sunday
         Jan 4 09:30:50 2017'
         name= Maindata[ :atpost].replace('From','').upper()
         print (name.replace('.',' ').upper().lstrip())
         print ("\n<<<<<<<<<<<<<<>>>>>>>>>>>>>")
         sppos=Maindata.find(' ', atpost)
         print (sppos)
         print (Maindata[ :sppos])
         host = Maindata [atpost + 1 : sppos ]
         print (host)
         print ("\n<<<<<<<<<<<<<<>>>>>>>>>>>>>")

OSSAMA EMBARAK
<<<<<<<<<<<<<<>>>>>>>>>>>>>
29
From ossama.embarak@hct.ac.ae
hct.ac.ae
<<<<<<<<<<<<<<>>>>>>>>>>>>>
```

Tabular Data and Data Formats

Data is available in different forms. It can be unstructured data, semistructured data, or structured data. Python provides different structures to maintain data and to manipulate it such as variables, lists, dictionaries, tuples, series, panels, and data frames. Tabular data can be easily represented in Python using lists of tuples representing the records of the data set in a data frame structure. Though easy to create, these kinds of representations typically do not enable important tabular data manipulations, such as efficient column selection, matrix mathematics, or spreadsheet-style operations. Tabular is a package of Python modules for working with tabular data. Its main object is the `tabarray` class, which is a data structure for holding and manipulating tabular data. You can put data into a `tabarray` object for more flexible and powerful data processing. The Pandas library also provides rich data structures and functions designed to make working with structured data fast, easy, and expressive. In addition, it provides a powerful and productive data analysis environment.

A Pandas data frame can be created using the following constructor:

```
pandas.DataFrame( data, index, columns, dtype, copy)
```

A Pandas data frame can be created using various input forms such as the following:

- List
- Dictionary
- Series
- Numpy ndarrays
- Another data frame

Chapter 3 will demonstrate the creation and manipulation of the data frame structure in detail.

Python Pandas Data Science Library

Pandas is an open source Python library providing high-performance data manipulation and analysis tools via its powerful data structures. The name Pandas is derived from "panel data," an econometrics term from multidimensional data. The following are the key features of the Pandas library:

- Provides a mechanism to load data objects from different formats
- Creates efficient data frame objects with default and customized indexing
- Reshapes and pivots date sets

- Provides efficient mechanisms to handle missing data
- Merges, groups by, aggregates, and transforms data
- Manipulates large data sets by implementing various functionalities such as slicing, indexing, subsetting, deletion, and insertion
- Provides efficient time series functionality

Sometimes you have to import the Pandas package since the standard Python distribution doesn't come bundled with the Pandas module. A lightweight alternative is to install Numpy using popular the Python package installer `pip`. The Pandas library is used to create and process series, data frames, and panels.

A Pandas Series

A *series* is a one-dimensional labeled array capable of holding data of any type (integer, string, float, Python objects, etc.). Listing 1-29 shows how to create a series using the Pandas library.

Listing 1-29. Creating a Series Using the Pandas Library

```
In [34]:#Create series from array using pandas and numpy
      import pandas as pd
      import numpy as np
      data = np.array([90,75,50,66])
      s = pd.Series(data,index=['A','B','C','D'])
      print (s)
A 90
B 75
C 50
D 66
dtype: int64
```

```
In [36]:print (s[1])

75

In [37]:#Create series from dictionary using pandas
      import pandas as pd
      import numpy as np
      data = {'Ahmed' : 92, 'Ali' : 55, 'Omar' : 83}
      s = pd.Series(data,index=['Ali','Ahmed','Omar'])
      print (s)

Ali 55
Ahmed 92
Omar 83
dtype: int64

In [38]:print (s[1:])

Ahmed 92
Omar 83
dtype: int64
```

A Pandas Data Frame

A *data frame* is a two-dimensional data structure. In other words, data is aligned in a tabular fashion in rows and columns. In the following table, you have two columns and three rows of data. Listing 1-30 shows how to create a data frame using the Pandas library.

Name	Age
Ahmed	35
Ali	17
Omar	25

Listing 1-30. Creating a Data Frame Using the Pandas Library

```
In [39]:import pandas as pd
      data = [['Ahmed',35],['Ali',17],['Omar',25]]
      DataFrame1 = pd.DataFrame(data,columns=['Name','Age'])
      print (DataFrame1)

   Name   Age
0  Ahmed  35
1  Ali    17
2  Omar   25
```

You can retrieve data from a data frame starting from index 1 up to the end of rows.

```
In [40]: DataFrame1[1:]

Out[40]:      Name   Age
         1    Ali    17
         2    Omar   25
```

You can create a data frame using a dictionary.

```
In [41]:import pandas as pd
      data = {'Name':['Ahmed', 'Ali', 'Omar',
      'Salwa'],'Age':[35,17,25,30]}
      dataframe2 = pd.DataFrame(data, index=[100, 101, 102, 103])
      print (dataframe2)

      Age    Name
100   35     Ahmed
101   17     Ali
102   25     Omar
103   30     Salwa
```

You can select only the first two rows in a data frame.

```
In [42]: dataframe2[:2]

Out[42]:      Age   Name
       100    35    Ahmed
       101    17    Ali
```

You can select only the name column in a data frame.

```
In [43]: dataframe2['Name']

Out[43]:100            Ahmed
101      Ali
102      Omar
103      Salwa
Name: Name, dtype: object
```

A Pandas Panels

A *panel* is a 3D container of data that can be created from different data structures such as from a dictionary of data frames, as shown in Listing 1-31.

Listing 1-31. Creating a Panel Using the Pandas Library

```
In [44]:# Creating a panel
      import pandas as pd
      import numpy as np
      data = {'Temperature Day1' : pd.DataFrame(np.random.
      randn(4, 3)),'Temperature Day2' : pd.DataFrame
      (np.random.randn(4, 2))}
      p = pd.Panel(data)
      print (p['Temperature Day1'])

        0          1          2
0   1.152400   -1.298529   1.440522
```

```
1   -1.404988   -0.105308   -0.192273
2   -0.575023   -0.424549   0.146086
3   -1.347784   1.153291   -0.131740
```

Python Lambdas and the Numpy Library

The lambda operator is a way to create small anonymous functions, in other words, functions without names. These functions are throwaway functions; they are just needed where they have been created. The lambda feature is useful mainly for Lisp programmers. Lambda functions are used in combination with the functions `filter()`, `map()`, and `reduce()`.

Anonymous functions refer to functions that aren't named and are created by using the keyword `lambda`. A lambda is created without using the `def` keyword; it takes any number of arguments and returns an evaluated expression, as shown in Listing 1-32.

Listing 1-32. Anonymous Function

```
In [34]:# Anonymous Function Definition
            summation=lambda val1, val2: val1 + val2#Call
            summation as a function
         print ("The summation of 7 + 10 = ", summation(7,10) )

The summation of 7 + 10 = 17

In [46]:result = lambda x, y : x * y
        result(2,5)

Out[46]: 10

In [47]:result(4,10)

Out[47]: 40
```

The map() Function

The map() function is used to apply a specific function on a sequence of data. The map() function has two arguments.

```
r = map(func, seq)
```

Here, func is the name of a function to apply, and seq is the sequence (e.g., a list) that applies the function func to all the elements of the sequence seq. It returns a new list with the elements changed by func, as shown in Listing 1-33.

Listing 1-33. Using the map() Function

```
In [65]:def fahrenheit(T):
              return ((float(9)/5)*T + 32)
           def celsius(T):
              return (float(5)/9)*(T-32)
        Temp = (15.8, 25, 30.5,25)
        F = list ( map(fahrenheit, Temp))
        C = list ( map(celsius, F))
        print (F)
        print (C)

[60.44, 77.0, 86.9, 77.0]
[15.799999999999999, 25.0, 30.500000000000004, 25.0]
```

```
In [72]:Celsius = [39.2, 36.5, 37.3, 37.8]
Fahrenheit = map(lambda x: (float(9)/5)*x + 32, Celsius)
for x in Fahrenheit:
      print(x)

102.56
97.7
99.14
100.03999999999999
```

The filter() Function

The `filter()` function is an elegant way to filter out all elements of a list for which the applied function returns true.

For instance, the function `filter(func, list1)` needs a function called `func` as its first argument. `func` returns a Boolean value, in other words, either true or false. This function will be applied to every element of the list `list1`. Only if `func` returns true will the element of the list be included in the result list.

The `filter()` function in Listing 1-34 is used to return only even values.

Listing 1-34. Using the filter() Function

```
In [79]:fib = [0,1,1,2,3,5,8,13,21,34,55]
        result = filter(lambda x: x % 2==0, fib)
        for x in result:
             print(x)

0
2
8
34
```

The reduce () Function

The `reduce()` function continually applies the function `func` to a sequence `seq` and returns a single value.

The `reduce()` function is used to find the max value in a sequence of integers, as shown in Listing 1-35.

Listing 1-35. Using the reduce() Function

```
In [81]: f = lambda a,b: a if (a > b) else b
reduce(f, [47,11,42,102,13])

102

In [82]: reduce(lambda x,y: x+y, [47,11,42,13])

113
```

Python Numpy Package

Numpy is a Python package that stands for "numerical Python." It is a library consisting of multidimensional array objects and a collection of routines for processing arrays.

The Numpy library is used to apply the following operations:

- Operations related to linear algebra and random number generation
- Mathematical and logical operations on arrays
- Fourier transforms and routines for shape manipulation

For instance, you can create arrays and perform various operations such as adding or subtracting arrays, as shown in Listing 1-36.

Listing 1-36. Example of the Numpy Function

```
In [83]:a=np.array([[1,2,3],[4,5,6]])
        b=np.array([[7,8,9],[10,11,12]])
        np.add(a,b)

Out[83]: array([[ 8, 10, 12], [14, 16, 18]])

In [84]:np.subtract(a,b) #Same as a-b

Out[84]: array([[-6, -6, -6], [-6, -6, -6]])
```

Data Cleaning and Manipulation Techniques

Keeping accurate data is highly important for any data scientist. Developing an accurate model and getting accurate predictions from the applied model depend on the missing values treatment. Therefore, handling missing data is important to make models more accurate and valid.

Numerous techniques and approaches are used to handle missing data such as the following:

- Fill NA forward
- Fill NA backward
- Drop missing values
- Replace missing (or) generic values
- Replace NaN with a scalar value

The following examples are used to handle the missing values in a tabular data set:

```
In [31]: dataset.fillna(0) # Fill missing values with zero value
In [35]: dataset.fillna(method='pad') # Fill methods Forward
In [35]: dataset.fillna(method=' bfill') # Fill methods Backward
In [37]: dataset.dropna() # remove all missing data
```

Chapter 5 covers different gathering and cleaning techniques.

Abstraction of the Series and Data Frame

A *series* is one of the main data structures in Pandas. It differs from lists and dictionaries. An easy way to visualize this is as two columns of data. The first is the special index, a lot like the dictionary keys, while the second is your actual data. You can determine an index for a series, or

Python can automatically assign indices. Different attributes can be used to retrieve data from a series' `iloc()` and `loc()` attributes. Also, Python can automatically retrieve data based on the passed value. If you pass an object, then Python considers that you want to use the index label-based `loc()`. However, if you pass an index integer parameter, then Python considers the `iloc()` attribute, as indicated in Listing 1-37.

Listing 1-37. Series Structure and Query

```
In [6]: import pandas as pd
         animals = ["Lion", "Tiger", "Bear"]
         pd.Series(animals)

Out[6]: 0 Lion
 1 Tiger
 2 Bear
dtype: object
```

You can create a series of numerical values.

```
In [5]: marks = [95, 84, 55, 75]
        pd.Series(marks)

Out[5]: 0   95
        1   84
        2   55
        3   75
        dtype: int64
```

You can create a series from a dictionary where indices are the dictionary keys.

```
In [11]: quiz1 = {"Ahmed":75, "Omar": 84, "Salwa": 70}
      q = pd.Series(quiz1)
      q
```

```
Out[11]: Ahmed    75
         Omar     84
         Salwa    70
         dtype: int64
```

The following examples demonstrate how to query a series.

You can query a series using a series label or the `lock()` attribute.

```
In [13]: q.loc['Ahmed']

Out[13]: 75

In [20]: q['Ahmed']


Out[20]: 75
```

You can query a series using a series index or the `ilock()` attribute.

```
In [19]: q.iloc[2]

Out[19]: 70

In [21]: q[2]

Out[21]: 70
```

You can implement a Numpy operation on a series.

```
In [25]:s = pd.Series([70,90,65,25, 99])
        s
Out[25]:0   70
        1   90
        2   65
        3   25
        4   99
        dtype: int64
```

```
In [27]:total =0
      for val in s:
             total += val
      print (total)
```

```
349
```

You can get faster results by using Numpy functions on a series.

```
In [28]: import numpy as np
          total = np.sum(s)
          print (total)
```

```
349
```

It is possible to alter a series to add new values; it is automatically detected by Python that the entered values are not in the series, and hence it adds it to the altered series.

```
In [29]:s = pd.Series ([99,55,66,88])
          s.loc['Ahmed'] = 85
          s
```

```
Out[29]: 0    99
         1    55
         2    66
         3    88
     Ahmed    85
     dtype: int64
```

You can append two or more series to generate a larger one, as shown here:

```
In [32]: test = [95, 84, 55, 75]
      marks = pd.Series(test)
      s = pd.Series ([99,55,66,88])
      s.loc['Ahmed'] = 85
```

```
NewSeries = s.append(marks)
NewSeries

Out[32]: 0    99
         1    55
         2    66
         3    88
     Ahmed    85
         0    95
         1    84
         2    55
         3    75
     dtype: int64
```

The *data frame* data structure is the main structure for data collection and processing in Python. A data frame is a two-dimensional series object, as shown in Figure 1-8, where there's an index and multiple columns of content each having a label.

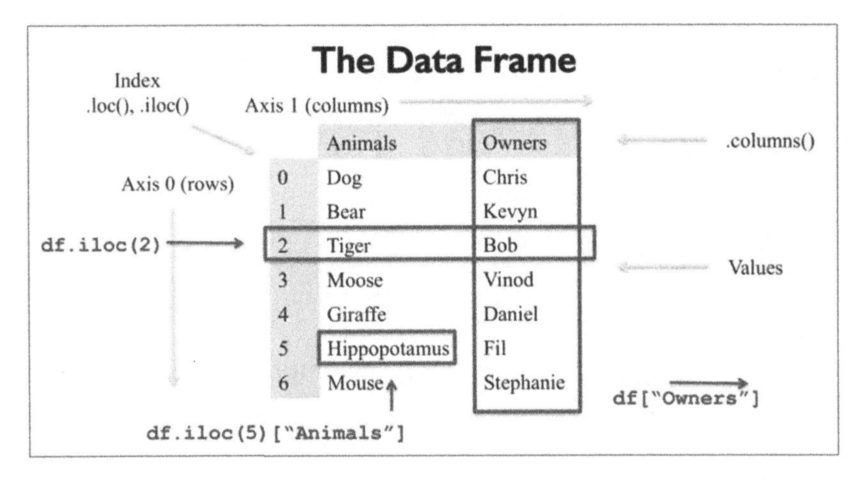

Figure 1-8. Data frame virtual structure

Data frame creation and queries were discussed earlier in this chapter and will be discussed again in the context of data collection structures in Chapter 3.

Running Basic Inferential Analyses

Python provides numerous libraries for inference and statistical analysis such as Pandas, SciPy, and Numpy. Python is an efficient tool for implementing numerous statistical data analysis operations such as the following:

- Linear regression
- Finding correlation
- Measuring central tendency
- Measuring variance
- Normal distribution
- Binomial distribution
- Poisson distribution
- Bernoulli distribution
- Calculating p-value
- Implementing a Chi-square test

Linear regression between two variables represents a straight line when plotted as a graph, where the exponent (power) of both of the variables is 1. A nonlinear relationship where the exponent of any variable is not equal to 1 creates a curve shape.

Let's use the built-in Tips data set available in the Seaborn Python library to find linear regression between a restaurant customer's total bill value and each bill's tip value, as shown in Figure 1-9. The function in Seaborn to find the linear regression relationship is `regplot`.

```
In [40]:import seaborn as sb
        from matplotlib import pyplot as plt
        df = sb.load_dataset('tips')
        sb.regplot(x = "total_bill", y = "tip", data = df)
        plt.xlabel('Total Bill')
        plt.ylabel('Bill Tips')
        plt.show()
```

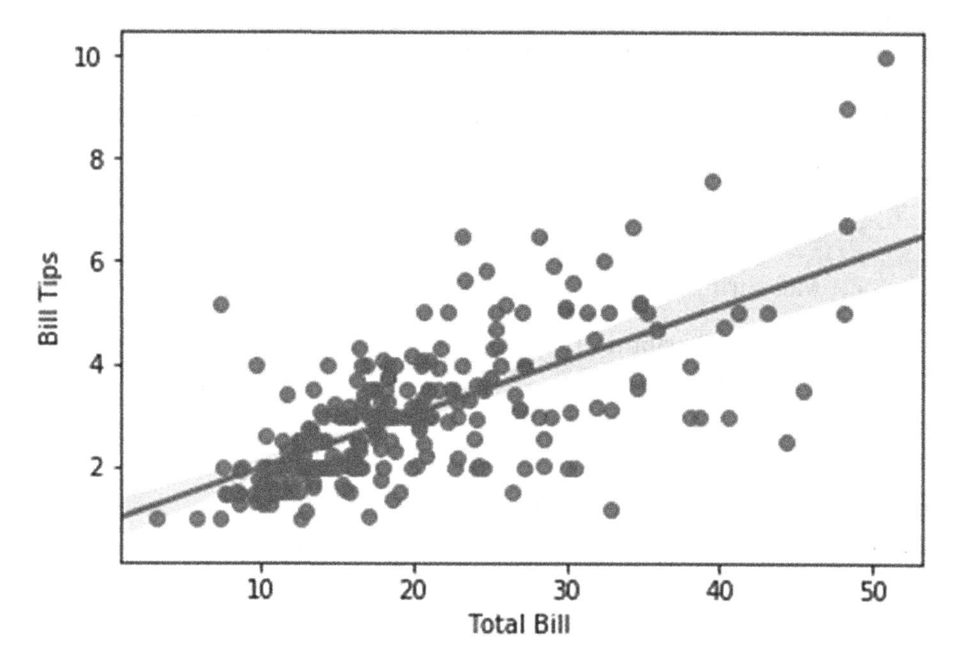

Figure 1-9. *Regression analysis*

Correlation refers to some statistical relationship involving dependence between two data sets, such as the correlation between the price of a product and its sales volume.

Let's use the built-in Iris data set available in the Seaborn Python library and try to measure the correlation between the length and the width of the sepals and petals of three species of iris, as shown in Figure 1-10.

```
In [42]: import matplotlib.pyplot as plt
          import seaborn as sns
          df = sns.load_dataset('iris')
          sns.pairplot(df, kind="scatter")
          plt.show()
```

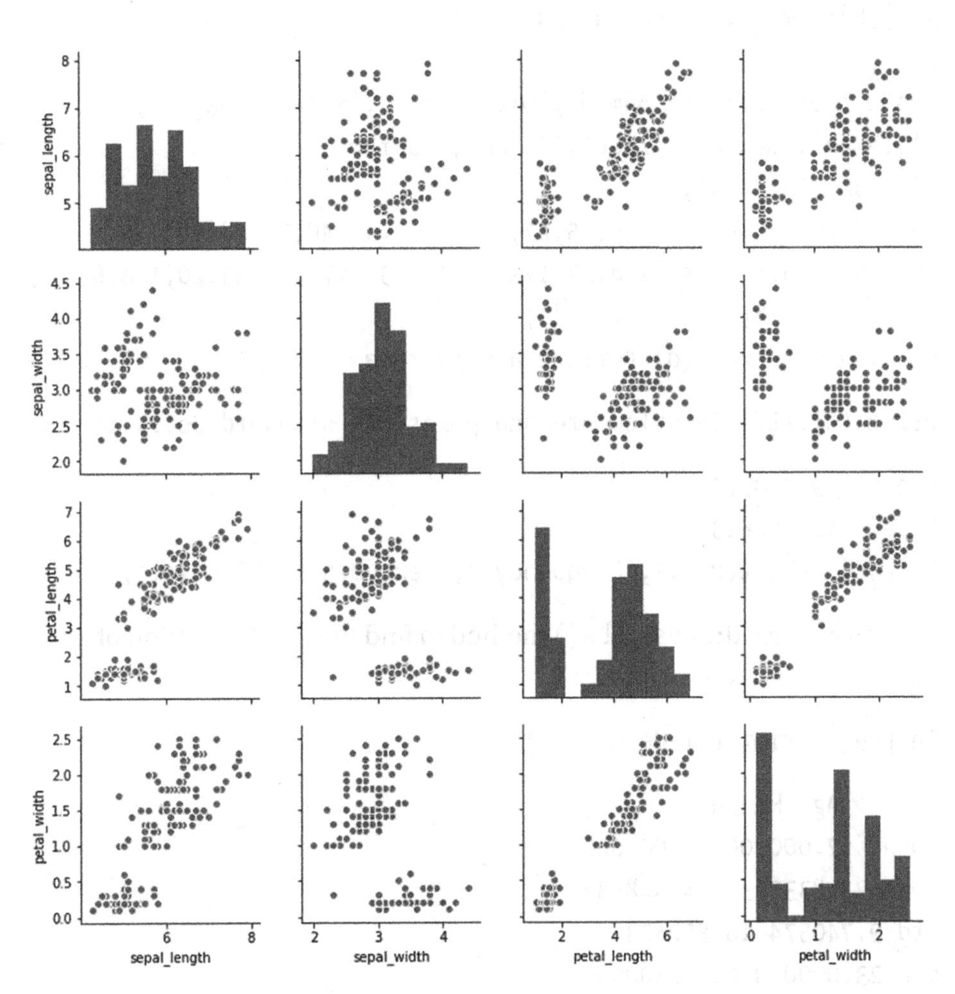

***Figure 1-10.** Correlation analysis*

In statistics, *variance* is a measure of how dispersed the values are from the mean value. Standard deviation is the square root of variance. In other words, it is the average of the squared difference of values in a data set from the mean value. In Python, you can calculate this value by using the function `std()` from the Pandas library.

```
In [58]: import pandas as pd
d = {
'Name': pd.Series(['Ahmed','Omar','Ali','Salwa','Majid',
 'Othman','Gameel','Ziad','Ahlam','Zahrah',
 'Ayman','Alaa']),
'Age': pd.Series([34,26,25,27,30,54,23,43,40,30,28,46]),
'Height':pd.Series([114.23,173.24,153.98,172.0,153.20,164.6,
 183.8,163.78,172.0,164.8 ])}
df = pd.DataFrame(d) #Create a DataFrame

print (df.std())# Calculate and print the standard deviation

Age      9.740574
Height 18.552823
Out[46]: [Text(0,0.5,'Frequency'), Text(0.5,0,'Binomial')]
```

You can use the `describe()` method to find the full description of a data frame set, as shown here:

```
In [59]: print (df.describe())

       Age Height
count 12.000000 12.000000
mean 33.833333 164.448333
std 9.740574 18.552823
min 23.000000 114.230000
25% 26.750000 161.330000
```

```
50% 30.000000 168.400000
75% 40.750000 173.455000
max 54.000000 183.800000
```

Central tendency measures the distribution of the location of values of a data set. It gives you an idea of the average value of the data in the data set and an indication of how widely the values are spread in the data set.

The following example finds the mean, median, and mode values of the previously created data frame:

```
In [60]: print ("Mean Values in the Distribution")
         print (df.mean())
         print ("*******************************")
         print ("Median Values in the Distribution")
         print (df.median())
         print ("*******************************")
         print ("Mode Values in the Distribution")
         print (df['Height'].mode())

Mean Values in the Distribution
Age 33.833333
Height     164.448333
dtype: float64
*******************************
Median Values in the Distribution
Age 30.0
Height     168.4
dtype: float64
*******************************
Mode Values of height in the Distribution
0     172.0
dtype: float64
```

Summary

This chapter introduced the data science field and the use of Python programming for implementation. Let's recap what was covered in this chapter.

- The data science main concepts and life cycle
- The importance of Python programming and its main libraries used for data science processing
- Different Python data structure use in data science applications
- How to apply basic Python programming techniques
- Initial implementation of abstract series and data frames as the main Python data structure
- Data cleaning and its manipulation techniques
- Running basic inferential statistical analyses

The next chapter will cover the importance of data visualization in business intelligence and much more.

Exercises and Answers

1. Write a Python script to prompt users to enter two values; then perform the basic arithmetical operations of addition, subtraction, multiplication, and division on the values.

Answer:

```
In [2]: # Store input numbers:
num1 = input('Enter first number: ')
```

```
num2 = input('Enter second number: ')
sumval = float(num1) + float(num2)     # Add two numbers
minval = float(num1) - float(num2)     # Subtract two numbers
mulval = float(num1) * float(num2)     # Multiply two numbers
divval = float(num1) / float(num2)     #Divide two numbers

# Display the sum
print('The sum of {0} and {1} is {2}'.format(num1, num2,
sumval))

# Display the subtraction
print('The subtraction of {0} and {1} is {2}'.format(num1, num2,
minval))

# Display the multiplication
print('The multiplication of {0} and {1} is {2}'.format(num1,
num2, mulval))

# Display the division
print('The division of {0} and {1} is {2}'.format(num1, num2,
divval))

Enter first number: 10
Enter second number: 5
The sum of 10 and 5 is 15.0
The subtraction of 10 and 5 is 5.0
The multiplication of 10 and 5 is 50.0
The division of 10 and 5 is 2.0
```

2. Write a Python script to prompt users to enter the lengths of a triangle sides. Then calculate the semiperimeters. Calculate the triangle area and display the result to the user. The area of a triangle is (s*(s-a)*(s-b)*(s-c))-1/2.

Answer:

```
In [3]:a = float(input('Enter first side: '))
       b = float(input('Enter second side: '))
       c = float(input('Enter third side: '))
       s = (a + b + c) / 2 # calculate the semiperimeter
       area = (s*(s-a)*(s-b)*(s-c)) ** 0.5 # calculate the area
       print('The area of the triangle is %0.2f' %area)

Enter first side: 10
Enter second side: 9
Enter third side: 7
The area of the triangle is 30.59
```

3. Write a Python script to prompt users to enter the first and last values and generate some random values between the two entered values.

Answer:

```
In [7]:import random
a = int(input('Enter the starting value : '))
b = int(input('Enter the end value : '))
print(random.randint(a,b))
random.sample(range(a, b), 3)

Enter the starting value : 10
Enter the end value : 100
14
Out[7]: [64, 12, 41]
```

4. Write a Python program to prompt users to enter a distance in kilometers; then convert kilometers to miles, where 1 kilometer is equal to 0.62137 miles. Display the result.

Answer:

```
In [9]: # convert kilometers to miles
kilometers = float(input('Enter the distance in kilometers: '))
# conversion factor
Miles = kilometers * 0.62137
print('%0.2f kilometers is equal to %0.2f miles'
 %(kilometers, Miles))

Enter the distance in kilometers: 120
120.00 kilometers is equal to 74.56 miles
```

5. Write a Python program to prompt users to enter a Celsius value; then convert Celsius to Fahrenheit, where T(°F) = T(°C) x 1.8 + 32. Display the result.

Answer:

```
In [11]: # convert Celsius to Fahrenheit
       Celsius = float(input('Enter temperature in Celsius: '))
       # conversion factor
       Fahrenheit = (Celsius * 1.8) + 32
       print('%0.2f Celsius is equal to %0.2f Fahrenheit'
 %(Celsius, Fahrenheit))

Enter temperature in Celsius: 25
25.00 Celsius is equal to 77.00 Fahrenheit
```

6. Write a program to prompt users to enter their working hours and rate per hour to calculate gross pay. The program should give the employee 1.5 times the hours worked above 30 hours. If Enter Hours is 50 and Enter Rate is 10, then the calculated payment is Pay: 550.0.

Answer:

```
In [6]:Hflage=True
      Rflage=True
      while Hflage & Rflage :
             hours = input ('Enter Hours:')
             try:
                   hours = int(hours)
                   Hflage=False
             except:
                   print ("Incorrect hours number !!!!")

             try:
                   rate = input ('Enter Rate:')
                   rate=float(rate)
                   Rflage=False
             except:
                    print ("Incorrect rate !!")

      if hours>40:
             pay= 40 * rate + (rate*1.5) * (hours - 40)
      else:
              pay= hours * rate
      print ('Pay:',pay)

Enter Hours: 50
Enter Rate: 10

Pay: 550.0
```

7. Write a program to prompt users to enter a value; then check whether the entered value is positive or negative value and display a proper message.

Answer:

```
In [1]: Val = float(input("Enter a number: "))
      if Val > 0:
             print("{0} is a positive number".format(Val))
      elif Val == 0:
             print("{0} is zero".format(Val))
      else:
             print("{0} is negative number".format(Val))

Enter a number: -12
-12.0 is negative number
```

8. Write a program to prompt users to enter a value; then check whether the entered value is odd or even and display a proper message.

Answer:

```
In [4]:# Check if a Number is Odd or Even
      val = int(input("Enter a number: "))
      if (val % 2) == 0:
             print("{0} is an Even number".format(val))
      else:
             print("{0} is an Odd number".format(val))

Enter a number: 13
13 is an Odd number
```

9. Write a program to prompt users to enter an age; then check whether each person is a child, a teenager, an adult, or a senior. Display a proper message.

Age	Category
< 13	Child
13 to 17	Teenager
18 to 59	Adult
> 59	Senior

Answer:

```
In [6]:age = int(input("Enter age of a person : "))
      if(age < 13):
             print("This is a child")
      elif(age >= 13 and age <=17):
             print("This is a teenager")
      elif(age >= 18 and age <=59):
             print("This is an adult")
      else:
             print("This is a senior")
```

```
Enter age of a person : 40

This is an adult
```

10. Write a program to prompt users to enter a car's speed; then calculate fines according to the following categories, and display a proper message.

Speed Limit	Fine Value
< 80	0
81 to 99	200
100 to 109	350
> 109	500

Answer:

```
In [7]:Speed = int(input("Enter your car speed"))
         if(Speed < 80):
             print("No Fines")
         elif(Speed >= 81 and Speed <=99):
             print("200 AE Fine ")
         elif(Speed >= 100 and Speed <=109):
             print("350 AE Fine ")
         else:
             print("500 AE Fine ")
```

```
Enter your car speed120

500 AE Fine
```

11. Write a program to prompt users to enter a year; then find whether it's a leap year. A year is considered a leap year if it's divisible by 4 and 100 *and* 400. If it's divisible by 4 and 100 but not by 400, it's not a leap year. Display a proper message.

Answer:

```
In [11]:year = int(input("Enter a year: "))
      if (year % 4) == 0:
            if (year % 100) == 0:
                  if (year % 400) == 0:
                        print("{0} is a leap year".
                        format(year))
                  else:
                        print("{0} is not a leap year".
                        format(year))
```

```
            else:
                  print("{0} is a leap year".format(year))
       else:
          print("{0} is not a leap year".format(year))
```

```
Enter a year: 2000

2000 is a leap year
```

12. Write a program to prompt users to enter a Fibonacci sequence. The Fibonacci sequence is the series of numbers 0, 1, 1, 2, 3, 5, 8, 13, 21, 34, ... The next number is found by adding the two numbers before it. For example, the 2 is found by adding the two numbers before it (1+1). Display a proper message.

Answer:

```
In [14]:nterms = int(input("How many terms you want? "))
             # first two terms
             n1 = 0
             n2 = 1
             count = 2
             # check if the number of terms is valid
             if nterms <= 0:
                   print("Please enter a positive integer")
             elif nterms == 1:
                   print("Fibonacci sequence:")
                   print(n1)
```

```
else:
      print("Fibonacci sequence:")
      print(n1,",",n2,end=', ') # end=', ' is used
      to continue printing in the same line
      while count < nterms:
            nth = n1 + n2
            print(nth,end=' , ')
            # update values
            n1 = n2
            n2 = nth
            count += 1
```

```
How many terms you want? 8

Fibonacci sequence:
0 , 1, 1 , 2 , 3 , 5 , 8 , 13 ,
```

CHAPTER 2

The Importance of Data Visualization in Business Intelligence

Data visualization is the process of interpreting data and presenting it in a pictorial or graphical format. Currently, we are living in the era of big data, where data has been described as a raw material for business. The volume of data used in businesses, industries, research organizations, and technological development is massive, and it is rapidly growing every day. The more data we collect and analyze, the more capable we can be in making critical business decisions. However, with the enormous growth of data, it has become harder for businesses to extract crucial information from the available data. That is where the importance of data visualization becomes clear. Data visualization helps people understand the significance of data by summarizing and presenting a huge amount of data in a simple and easy-to-understand format in order to communicate the information clearly and effectively.

O. Embarak, *Data Analysis and Visualization Using Python*,
https://doi.org/10.1007/978-1-4842-4109-7_2

Shifting from Input to Output

A decision-maker for any business wants to access highly visual business intelligence (BI) tools that can help to make the right decisions quickly. Business intelligence has become more mainstream; hence, vendors are beginning to focus on both ends of the pipeline and improve the quality of data input. There is also a strong focus on ensuring that the output is well-structured and clearly presented. This focus on output has largely been driven by the demands of consumers, who have been enticed by what visualization can offer. A BI dashboard can be a great way to compile several different data visualizations to provide an at-a-glance overview of business performance and areas for improvement.

Why Is Data Visualization Important?

A picture is worth a thousand words, as they say. Humans just understand data better through pictures rather than by reading numbers in rows and columns. Accordingly, if the data is presented in a graphical format, people are more able to effectively find correlations and raise important questions.

Data visualization helps the business to achieve numerous goals.

- Converting the business data into interactive graphs for dynamic interpretation to serve the business goals
- Transforming data into visually appealing, interactive dashboards of various data sources to serve the business with the insights
- Creating more attractive and informative dashboards of various graphical data representations
- Making appropriate decisions by drilling into the data and finding the insights

- Figuring out the patterns, trends, and correlations in the data being analyzed to determine where they must improve their operational processes and thereby grow their business
- Giving a fuller picture of the data under analysis
- Organizing and presenting massive data intuitively to present important findings from the data
- Making better, quick, and informed decisions with data visualization

Why Do Modern Businesses Need Data Visualization?

With the huge volume of data collected about business activities using different means, business leaders need proper techniques to easily drill down into the data to see where they can improve operational processes and grow their business. Data visualization brings business intelligence to reality. Data visualization is needed by modern businesses for these reasons:

- Data visualization helps companies to analyze its different processes so the management can focus on the areas for improvement to generate more revenue and improve productivity.
- It brings business intelligence to life.
- It applies a creative approach to understanding the hidden information within the business data.
- It provides a better and faster way to identify patterns, trends, and correlation in the data sets that would remain undetected with just text.

- It identifies new business opportunities by predicting upcoming trends or sales volumes and the revenue they will generate.
- It supplies managers with information they need to make more effective comparisons between data sets by plotting them on the same visualization.
- It enables managers to understand the correlations between the operating conditions and the business performance.
- It helps businesses to discover the gray areas of the business and make the right decisions for improvement.
- Data visualization helps managers to understand customers' behaviors and interests and hence retains customers and market share.

The Future of Data Visualization

Data visualization is moving from being an art to being a science field. Data science technologies impose the need to move from relatively simple graphs to multifaceted relational maps. Multidimensional visualizations will boost the role that data visualizations can play in the Internet of Things, network and complexity theories, nanoscience, social science research, education systems, conative science, space, and much more. Data visualization will play a vital role, now and in the future, in applying many concepts such as network theory, Internet of Things, complexity theory, and more. For instance, network theory employs algorithms to understand and model pair-wise relationships between objects to understand relationships and interactions in a variety of domains, such as crime prevention and disease management, social network analysis, biological network analysis, network optimization, and link analysis.

Data visualization will be used intensively to analyze and visualize data streams collected from billions of interconnected devices, from smart appliances and wearables to automobile sensors and environmental and smart cities monitors. Internet of Things device data will provide extraordinary insight into what's happening around the globe. In this context, data visualization will improve safety levels, drive operational efficiencies, help to better understand several worldwide phenomena, and improve and customize provided intercontinental services.

How Data Visualization Is Used for Business Decision-Making

Data visualization is a real asset for any business to help make real-time business decisions. It visualizes extracted information into logical and meaningful parts and helps users avoid information overload by keeping things simple, relevant, and clear. There are many ways in which visualizations help a business to improve its decision-making.

Faster Responses

Quick response to customers' or users' requirements is important for any company to retain their clients, as well as to keep their loyalty. With the massive amount of data collected daily via social networks or via companies' systems, it becomes incredibly useful to put useful interpretations of the collected data into the hands of managers and decision-makers so they can quickly identify issues and improve response times.

Simplicity

It is impossible to make efficient decisions based on large amounts of raw data. Therefore, data visualization gives the full picture of the scoped parameters and simplifies the data by enabling decision-makers to cherry-pick the relevant data they need and dive into a detailed view wherever is needed.

Easier Pattern Visualization

Data visualization provides easier approaches to identifying upcoming trends and patterns within data sets and hence enables businesses to make efficient decisions and prepare strategies in advance.

Team Involvement

Data visualizations process not only historical data but also real-time data. Different organization units gain the benefit of having direct access to the extracted information displayed by data visualization tools. This increases the levels of collaboration between departments to help them achieve strategic goals.

Unify Interpretation

Data visualizations can produce charts and graphics that lead to the same interpretations by all who use the extracted information for decision-making. There are many data visualization tools such as R, Python, Matlab, Scala, and Java. Table 2-1 compares the most common languages, which are the R and Python languages.

Table 2-1. The R Language vs. Python

Parameter	R	Python
Main use	Data analysis and statistics.	Deployment and production.
Users	Scholars and researchers.	Programmers and developers.
Flexibility	Easy-to-use available library.	It's easy to construct new models from scratch.
Integration	Runs locally.	Well-integrated with app. Runs through the cloud.
Database size	Handles huge size.	Handles huge size.
IDE examples	RStudio.	Spyder, IPython Notebook, Jupyter Notebook, etc.
Important packages and libraries	Tydiverse, Ggplot2, Caret, Zoo.	Pandas, Numpy, Scipy, Scikit-learn, TensorFlow, Caret.
Advantages	• Comprehensive statistical analysis package. • Open source; anyone can use it. • It is cross-platform and can run on many operating systems. • Anyone can fix bugs and make code enhancements.	• Python is a general-purpose language that is easy and intuitive. • Useful for mathematical computation. • Can share data online via clouds and IDEs such as Jupyter Notebook. • Can be deployed. • Fast processing. • High code readability. • Supports multiple systems and platforms. • Easy integration with other languages such as C and Java.

(*continued*)

Table 2-1. *(continued)*

Parameter	R	Python
Disadvantages	• Quality of some packages is not good. • R can consume all the memory because of its memory management. • Slow and high learning curve. • Dependencies between library. • There is no regular and direct update for R packages and bugs.	• Comparatively smaller pool of Python developers. • Python doesn't have as many libraries as R. • Not good for mobile development. • Database access limitations.

Introducing Data Visualization Techniques

Data visualization aims to understand data by extracting and graphing information to show patterns, spot trends, and identify outliers. There are two basic types of data visualization.

- *Exploration* helps to extract information from the collected data.
- *Explanation* demonstrates the extracted information.

There are many types of 2D data visualizations, such as temporal, multidimensional, hierarchical, and network. In the following section, we demonstrate numerous data visualization techniques provided by the Python programming language.

Loading Libraries

Some libraries are bundled with Python, while others should be directly downloaded and installed.

For instance, you can install Matplotlib using pip as follows:

```
python -m pip install -U pip setuptools
python -m pip install matplotlib
```

You can install, search, or update Python packages with Jupyter Notebook or with a desktop Python IDE such as Spyder. Table 2-2 shows how to use the pip and conda commands.

Table 2-2. *Installing and Upgrading Python Packages*

Description	pip	conda Anaconda
Works with	Python and Anaconda	Anaconda only
Search a package	pip search matplolib	conda search matplolib
Install a package	pip install matplolib	conda install matplolib
Upgrade a package	pip install matplolib-upgrade	conda install matplolib-upgrade
Display installed packages	pip list	conda list

Let's list all the installed or upgraded Python libraries using the pip and conda commands.

```
conda list
```

```
pip list
```

Similarly, you can install or upgrade packages or specific Python packages such as Matplotlib on Jupyter Notebooks, as shown in Listing 2-1.

Listing 2-1. Installed or Upgraded Packages

```
In [5]: try:
      import matplotlib
            except:
             import pip pip.main(['install', 'matplotlib'])
             import matplotlib
```

It is possible to import any library and use alias names, as shown here:

```
In [ ]:import matplotlib.pyplot as plt import numpy as np
      import pandas as pd
      import seaborn as sns
      import pygal from mayavi
      import mlab
      etc....
```

Once you load any library to your Python script, then you can call the package functions and attributes.

Popular Libraries for Data Visualization in Python

The Python language provides numerous data visualization libraries for plotting data. The most used and common data visualization libraries are Pygal, Altair, VisPy, PyQtGraph, Matplotlib, Bokeh, Seaborn, Plotly, and ggplot, as shown in Figure 2-1.

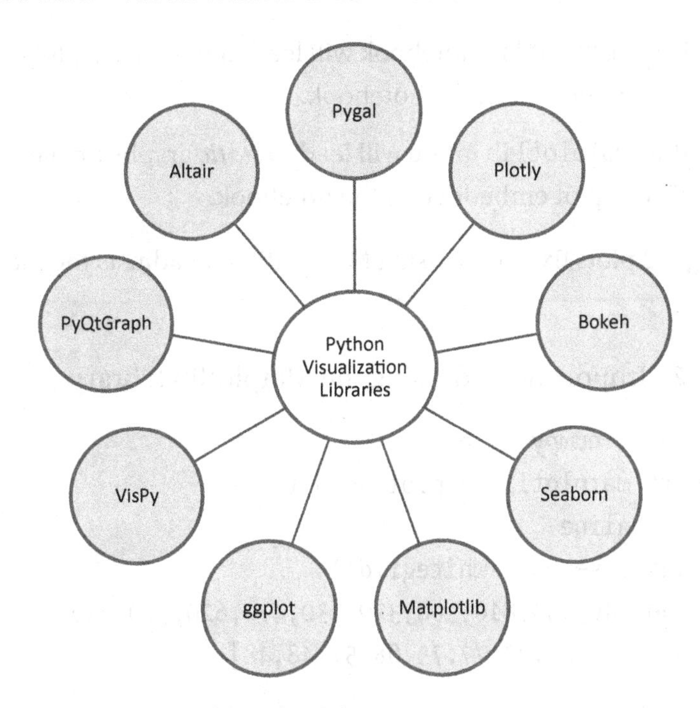

Figure 2-1. *Data visualization libraries*

Each of these libraries has its own features. Some of these libraries may be adopted for implementation and dependent on other libraries. For example, Seaborn is a statistical data visualization library that uses Matplotlib. In addition, it needs Pandas and maybe NumPy for statistical processing before visualizing data.

Matplotlib

Matplotlib is a Python 2D plotting library for data visualization built on Numpy arrays and designed to work with the broader SciPy stack. It produces publication-quality figures in a variety of formats and interactive environments across platforms. There are two options for embedding graphics directly in a notebook.

- The `%matplotlib` notebook will lead to *interactive* plots embedded within the notebook.
- The `%matplotlib` inline will lead to *static* graphs images of your plot embedded in the notebook.

Listing 2-2 plots fixed data using Matplotlib and adjusts the plot attributes.

Listing 2-2. Importing and Using the Matplotlib Library

```
In [12]:import numpy as np
      import matplotlib.pyplot as plt
%matplotlib inline
plt.style.use('seaborn-whitegrid')
X = [590,540,740,130,810,300,320,230,470,620,770,250]
Y = [32,36,39,52,61,72,77,75,68,57,48,48]

plt.scatter(X,Y)
plt.xlim(0,1000)
plt.ylim(0,100)

#scatter plot color
plt.scatter(X, Y, s=60, c='red', marker='^')

#change axes ranges
plt.xlim(0,1000)
plt.ylim(0,100)

#add title
plt.title('Relationship Between Temperature and Iced
Coffee Sales')

#add x and y labels
plt.xlabel('Sold Coffee')
plt.ylabel('Temperature in Fahrenheit')
```

```
#show plot
plt.show()
```

Figure 2-2 shows a visualization in the Matplot library.

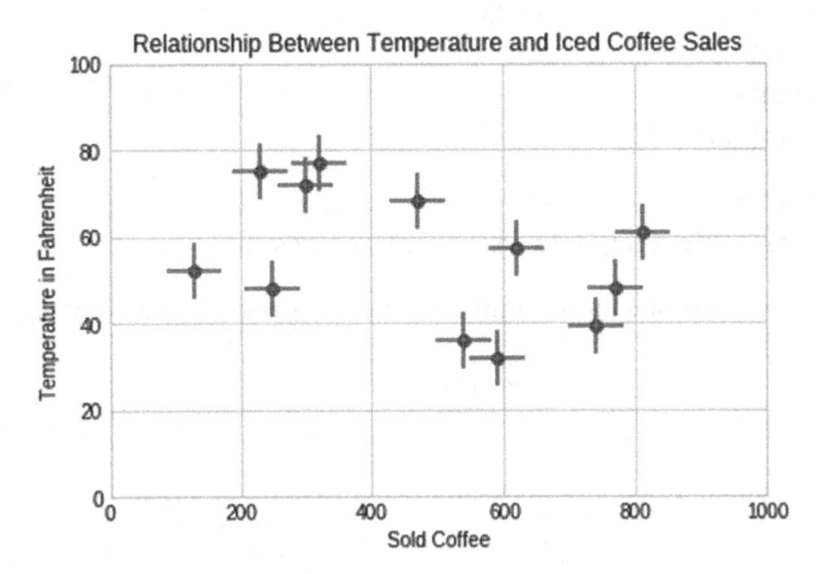

Figure 2-2. *Visualizing data using Matplotlib*

Listing 2-3 plots fixed data using Matplotlib and adjusts the plot attributes.

Listing 2-3. Importing Numpy and Calling Its Functions

```
In [20]:%matplotlib inline
      import matplotlib.pyplot as plt
import numpy as np
plt.style.use('seaborn-whitegrid')

# Create empty figure
fig = plt.figure()
ax = plt.axes()
x = np.linspace(0, 10, 1000)
```

```
ax.plot(x, np.sin(x));
plt.plot(x, np.sin(x))
plt.plot(x, np.cos(x))

# set the x and y axis range
plt.xlim(0, 11)
plt.ylim(-2, 2)
plt.axis('tight')

#add title
plt.title('Plotting data using sin and cos')
```

Figure 2-3 shows the accumulated attributes added to the same graph.

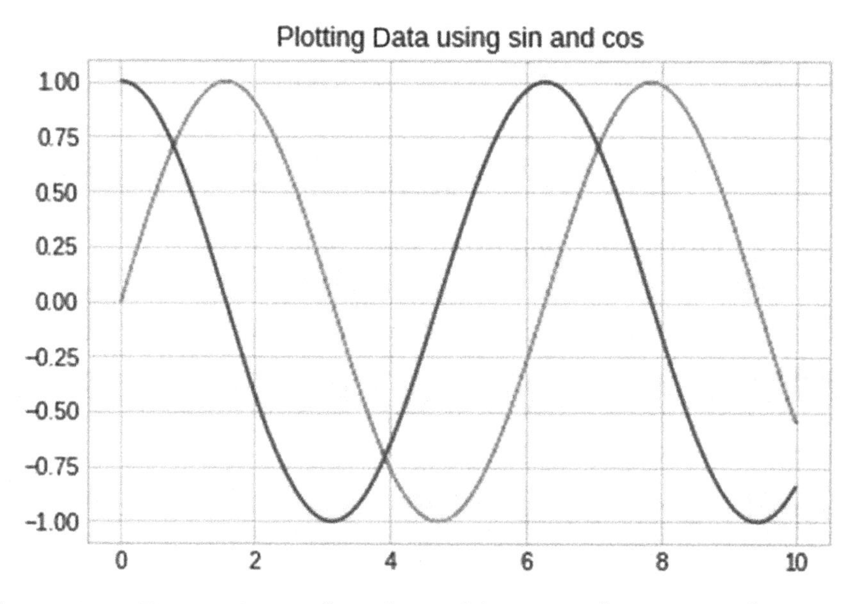

***Figure 2-3.** Determining the adapted function (sin and cos) by Matplotlib*

All altered attributes are applied to the same graph as shown above.

There are many different plotting formats generated by the Matplotlib package; some of these formats will be discussed in Chapter 7.

Seaborn

Seaborn is a Python data visualization library based on Matplotlib that provides a high-level interface for drawing attractive and informative statistical graphics (see Listing 2-4).

Listing 2-4. Importing and Using the Seaborn Library

```
In [34]: import matplotlib.pyplot as plt
%matplotlib inline
import numpy as np
import pandas as pd
import seaborn as sns
plt.style.use('classic')
plt.style.use('seaborn-whitegrid')
# Create some data
data = np.random.multivariate_normal([0, 0], [[5, 2], [2, 2]],
size=2000)
data = pd.DataFrame(data, columns=['x', 'y'])
# Plot the data with seaborn
sns.distplot(data['x'])
sns.distplot(data['y']);
```

Figure 2-4 shows a Seaborn graph.

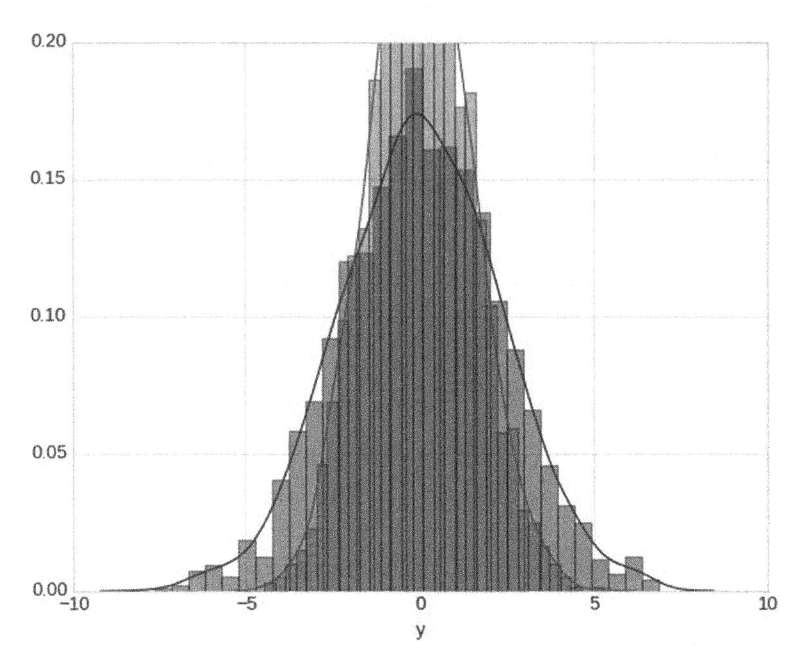

Figure 2-4. *Seaborn graph*

Let's use the distribution using a kernel density estimation, which Seaborn does with `sns.kdeplot`. You can use the same data set, called Data, as in the previous example (see Figure 2-5).

```
In [35]: for col in 'xy':
      sns.kdeplot(data[col], shade=True)
```

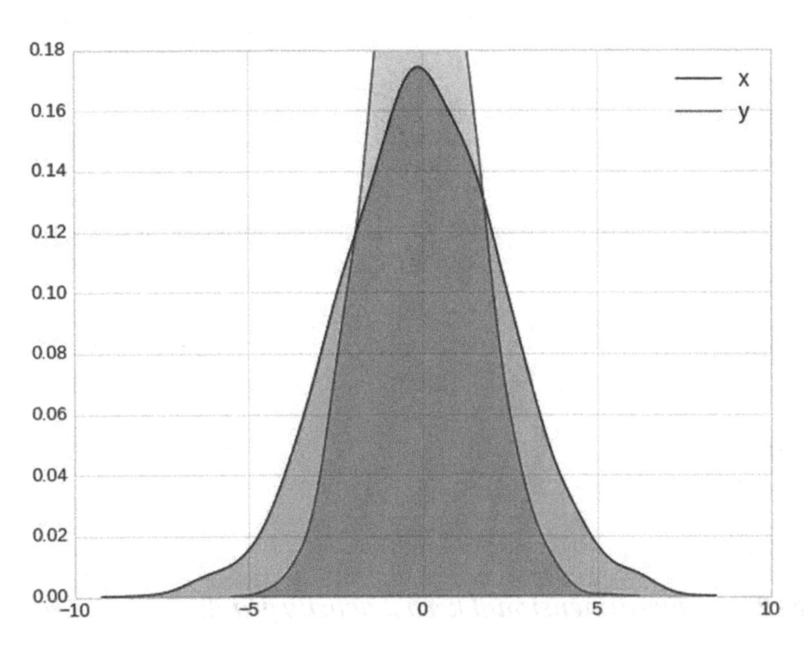

Figure 2-5. *Seaborn kernel density estimation graph*

Passing the full two-dimensional data set to kdeplot as follows, you will get a two-dimensional visualization of the data (see Figure 2-6):

```
In [36]: sns.kdeplot(data);
```

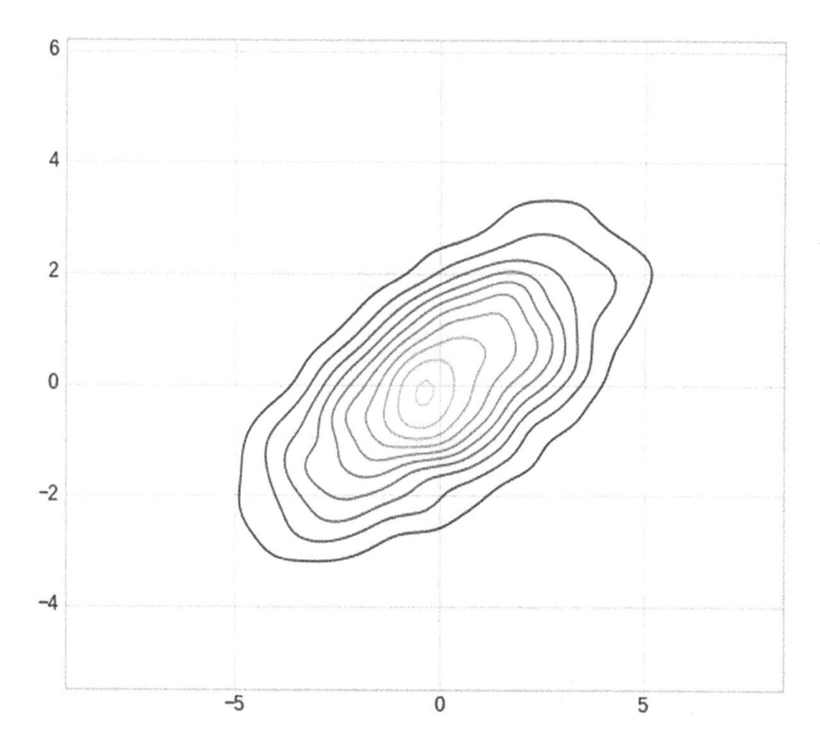

Figure 2-6. *Two-dimensional kernel density graph*

Let's use the joint distribution and the marginal distributions together using `sns.jointplot`, as shown here (see Figure 2-7):

```
In [37]:      with sns.axes_style('white'):
      sns.jointplot("x", "y", data, kind='kde');
```

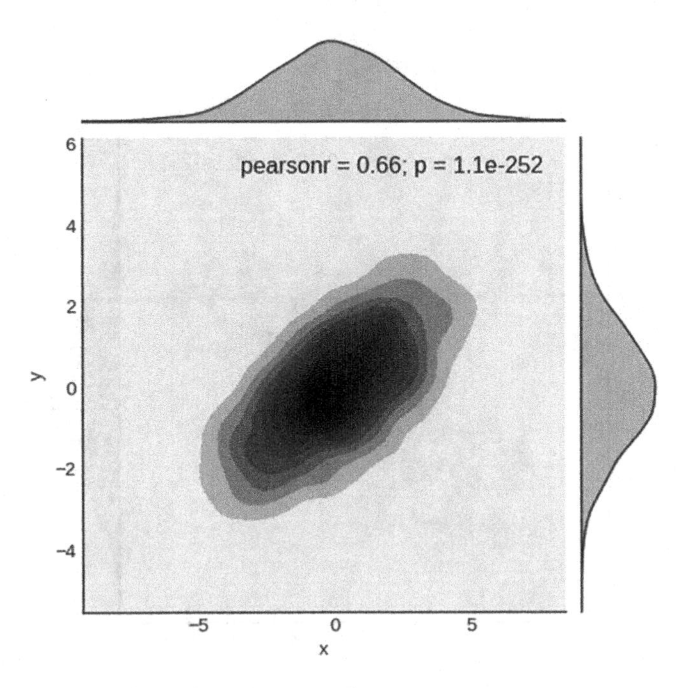

Figure 2-7. *Joint distribution graph*

Use a hexagonally based histogram in the joint plot, as shown here (see Figure 2-8):

```
In [38]:     with sns.axes_style('white'):
      sns.jointplot("x", "y", data, kind='hex')
```

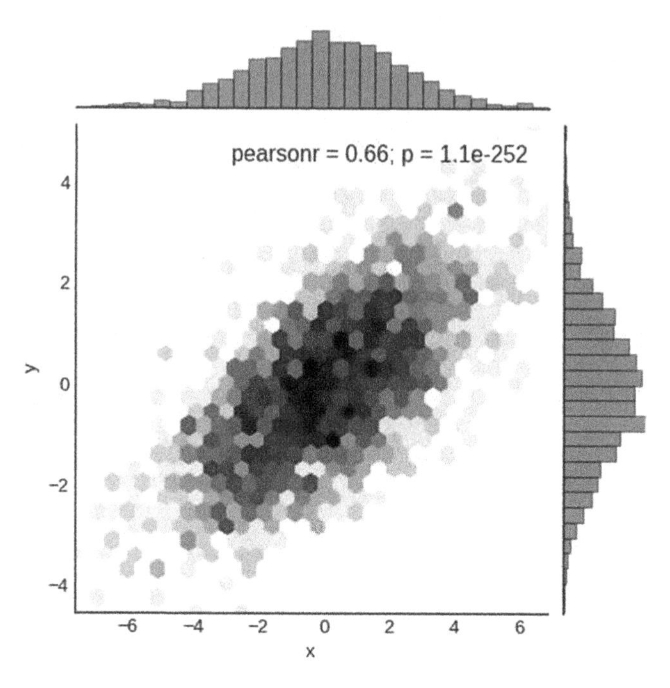

Figure 2-8. *A hexagonally based histogram graph*

You can also visualize multidimensional relationships among the samples by calling `sns.pairplot` (see Figure 2-9):

```
In [41]: sns.pairplot(data);
```

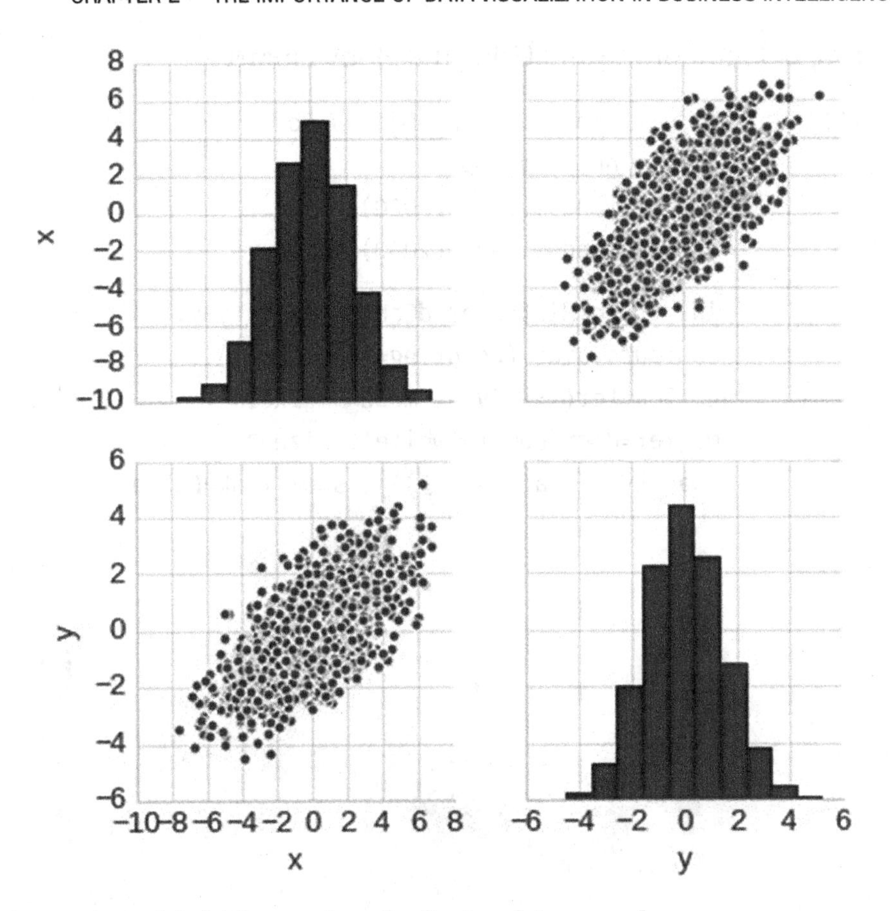

Figure 2-9. *Multidimensional relationships graph*

There are many different plotting formats generated by the Seaborn package; some of these formats will be discussed in Chapter 7.

Plotly

The Plotly Python graphing library makes interactive, publication-quality graphs online. Different dynamic graphs formats can be generated online or offline.

Listing 2-5 implements a dynamic heatmap graph (see Figure 2-10).

Listing 2-5. Importing and Using the Plotly Library

```
In [67]:    import plotly.graph_objs as go
            import numpy as np
            x = np.random.randn(2000)
            y = np.random.randn(2000)

            iplot([go.Histogram2dContour(x=x, y=y,
            contours=dict (coloring='heatmap')),
            go.Scatter(x=x, y=y, mode='markers',
            marker=dict(color='white', size=3,
            opacity=  opacity=0.3))], show_link=False)
```

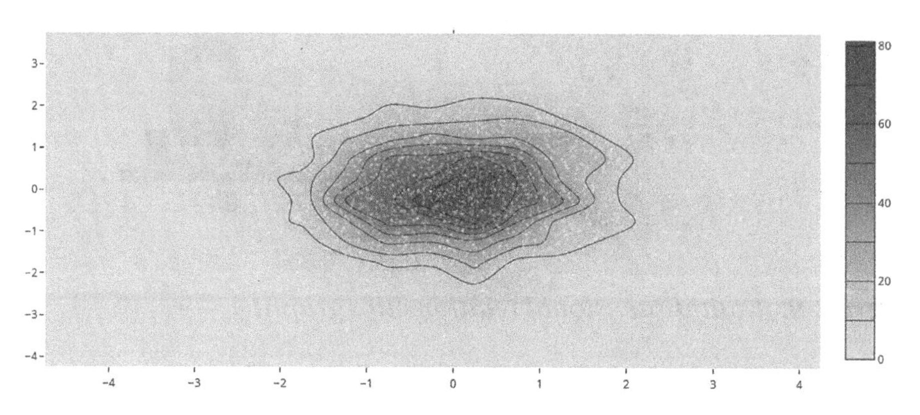

Figure 2-10. *Dynamic heatmap graph*

Use `plotly.offline` to execute the Plotly script offline within a notebook (Figure 2-11), as shown here:

```
In [90]:    import plotly.offline as offline
                import plotly.graph_objs as go
                offline.plot({'data': [{'y': [14, 22, 30,
                44]}],
```

```
                    'layout': {'title': 'Offline Plotly', 'font':
                    dict(size=16)}}, image='png')
Out[90]: 'file:///home/nbuser/library/temp-plot.html'
```

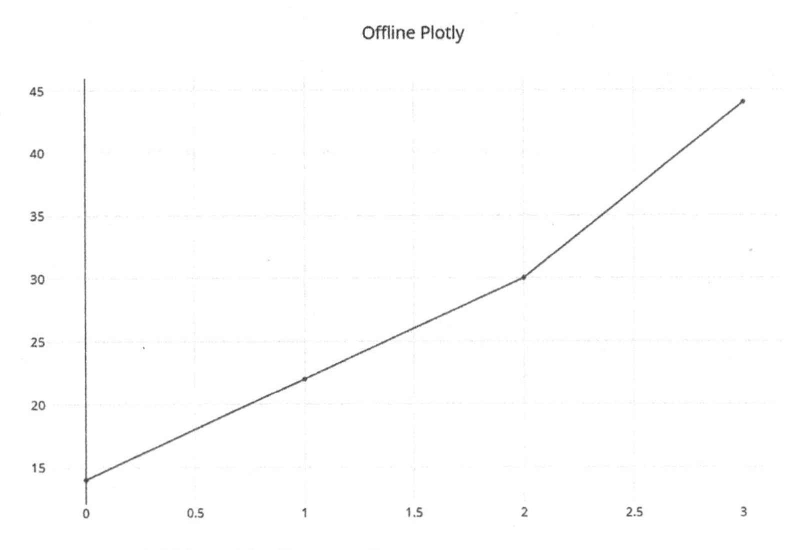

Figure 2-11. *Offline Plotly graph*

Executing the Plotly Python script, as shown in Listing 2-6, will open a web browser with the dynamic Plotly graph drawn, as shown in Figure 2-12.

Listing 2-6. Importing and Using the Plotly Package

```
In [64]:from plotly import __version__
     from plotly.offline import download_plotlyjs,
     init_notebook_mode, plot, iplot init_notebook_
     mode(connected=True)
     print (__version__)

<inline script removed for security reasons>
3.1.0
```

```
In [91]: import plotly.graph_objs as go
      plot([go.Scatter(x=[95, 77, 84], y=[75, 67, 56])])
Out[91]: 'file:///home/nbuser/library/temp-plot.html'
```

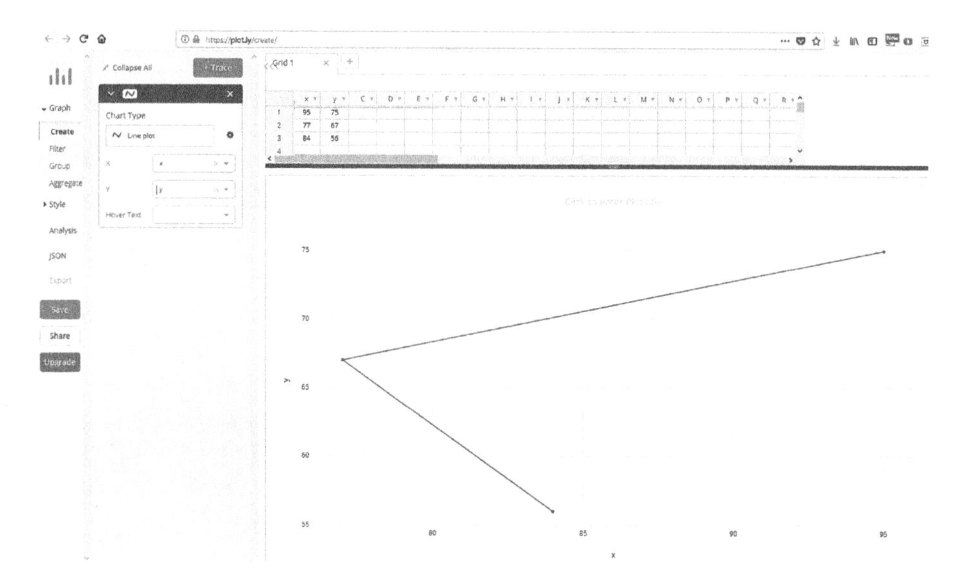

Figure 2-12. *Plotly dynamic graph*

Plotly graphs are more suited to dynamic and online data visualization, especially for real-time data streaming, which isn't covered in this book.

Geoplotlib

Geoplotlib is a toolbox for creating a variety of map types and plotting geographical data. Geoplotlib needs Pyglet as an object-oriented programming interface. This type of plotting is not covered in this book.

Pandas

Pandas is a Python library written for data manipulation and analysis. You can use Python with Pandas in a variety of academic and commercial domains, including finance, economics, statistics, advertising, web analytics, and much more. Pandas is covered in Chapter 6.

Introducing Plots in Python

As indicated earlier, numerous plotting formats can be used, even offline or online ones. The following are examples of direct plotting.

Listing 2-7 implements a basic plotting plot. Figure 2-13 shows the graph.

Listing 2-7. Running Basic Plotting

```
In [116]: import pandas as pd import numpy as np
 df = pd.DataFrame(np.random.randn(200,6),index= pd.date_
range('1/9/2009', periods=200), columns= list('ABCDEF'))
df.plot(figsize=(20, 10)).legend(bbox_to_anchor=(1, 1))
```

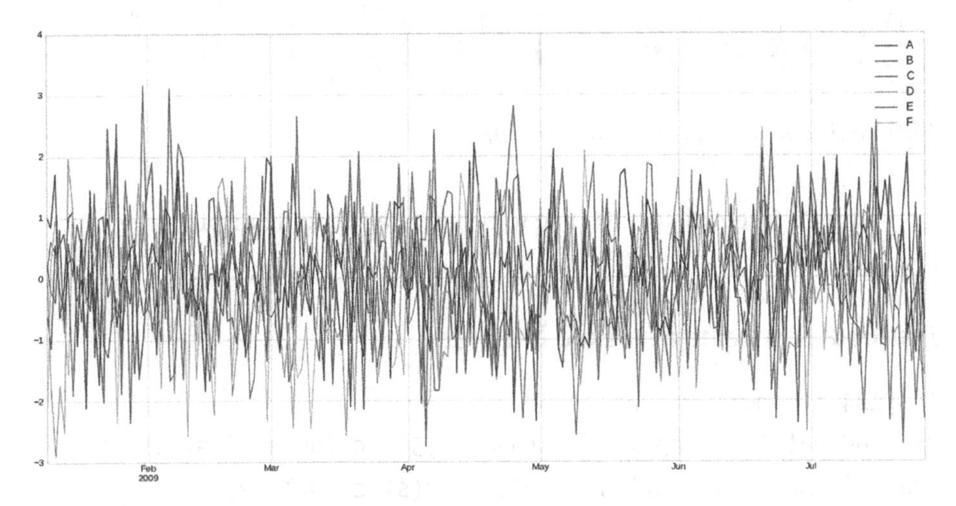

Figure 2-13. *Direct plot graph*

Listing 2-8 creates a bar plot graph (see Figure 2-14).

Listing 2-8. Direct Plotting

```
In [123]: import pandas as pd
          import numpy as np
```

```
df = pd.DataFrame(np.random.rand(20,5), columns=['Jan','Feb',
'March','April', 'May'])
df.plot.bar(figsize=(20, 10)).legend(bbox_to_anchor=(1.1, 1))
```

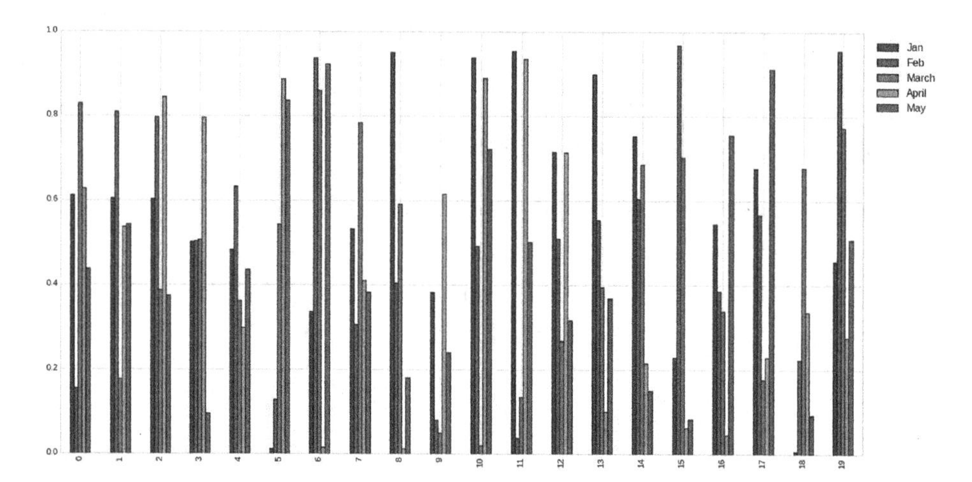

Figure 2-14. *Direct bar plot graph*

Listing 2-9 sets `stacked=True` to produce a stacked bar plot (see Figure 2-15).

Listing 2-9. Create a stacked bar plot

```
In [124]: import pandas as pd
df = pd.DataFrame(np.random.rand(20,5), columns=['Jan','Feb',
'March','April', 'May']) df.plot.bar(stacked=True,
figsize=(20, 10)).legend(bbox_to_anchor=(1.1, 1))
```

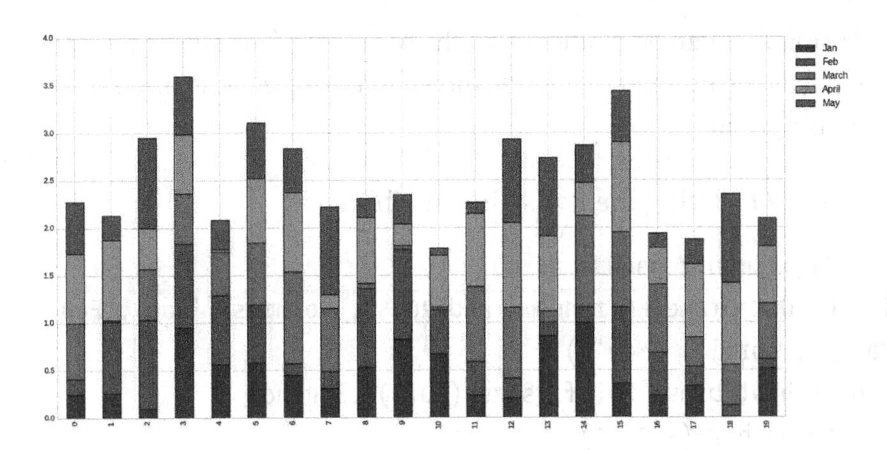

Figure 2-15. *Stacked bar plot graph*

To get horizontal bar plots, use the barh method, as shown in Listing 2-10. Figure 2-16 shows the resulting graph.

Listing 2-10. Bar Plots

```
In [126]: import pandas as pd
df = pd.DataFrame(np.random.rand(20,5), columns=['Jan','Feb',
'March','April', 'May']) df.plot.barh(stacked=True,
figsize=(20, 10)).legend(bbox_to_anchor=(1.1, 1))
```

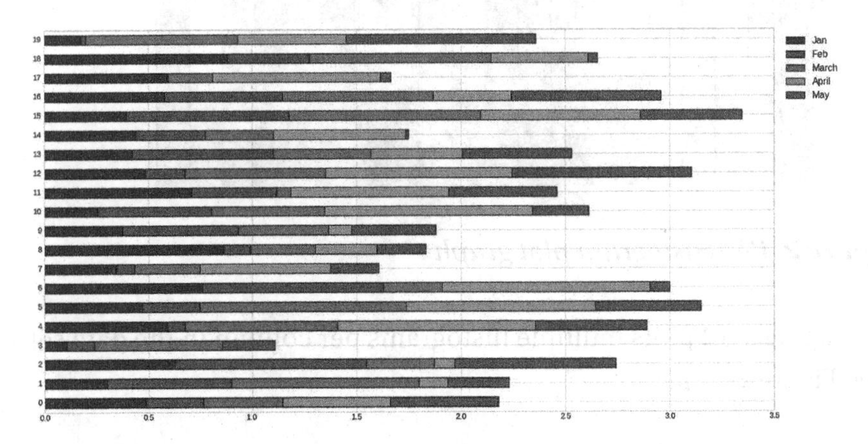

Figure 2-16. *Horizontal bar plot graph*

Histograms can be plotted using the `plot.hist()` method; you can also specify the number of bins, as shown in Listing 2-11. Figure 2-17 shows the graph.

Listing 2-11. Using the Bar's bins Attribute

```
In [131]: import pandas as pd
df = pd.DataFrame(np.random.rand(20,5), columns=['Jan','Feb',
'March','April', 'May'])
df.plot.hist(bins= 20, figsize=(10,8)).legend
bbox_to_anchor=(1.2, 1))
```

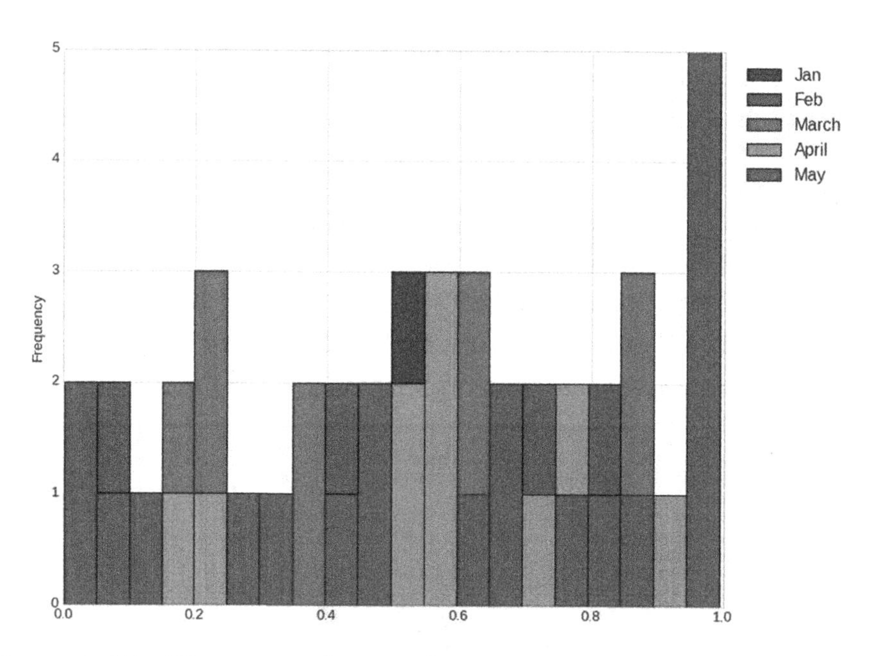

Figure 2-17. *Histogram plot graph*

Listing 2-12 plots multiple histograms per column in the data set (see Figure 2-18).

Listing 2-12. Multiple Histograms per Column

```
In [139]: import pandas as pd
          import numpy as np
df=pd.DataFrame({'April':np.random.randn(1000)+1,'May':np.random.
randn(1000),'June': np.random.randn(1000) - 1}, columns=['April',
'May', 'June'])
df.hist(bins=20)
```

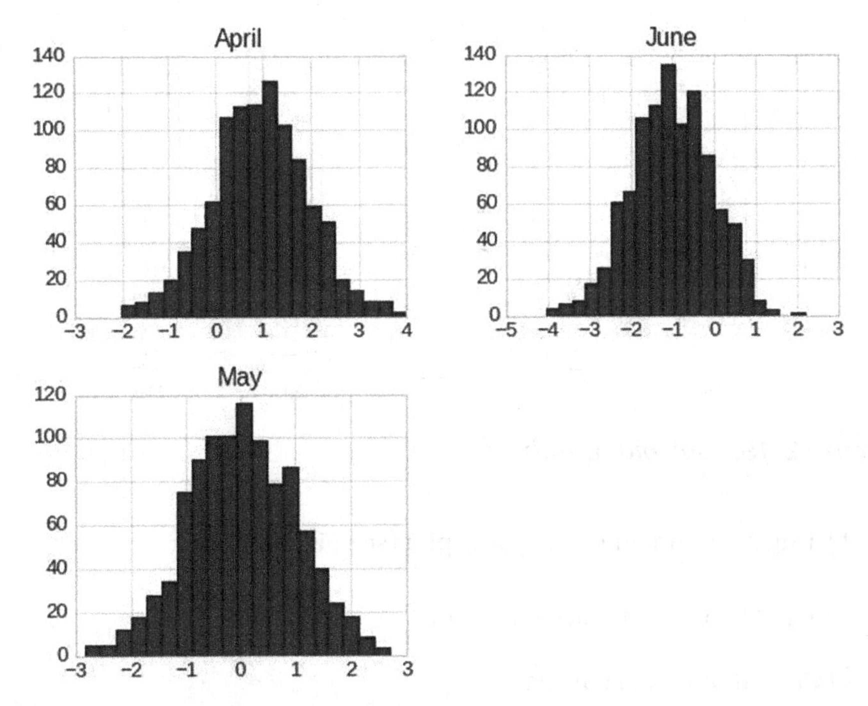

Figure 2-18. *Column base histograms plot graph*

Listing 2-13 implements a box plot (see Figure 2-19).

Listing 2-13. Creating a Box Plot

```
In [140]:import pandas as pd
           import numpy as np
```

```
df = pd.DataFrame(np.random.rand(20,5),
     columns=['Jan','Feb','March','April', 'May'])
df.plot.box()
```

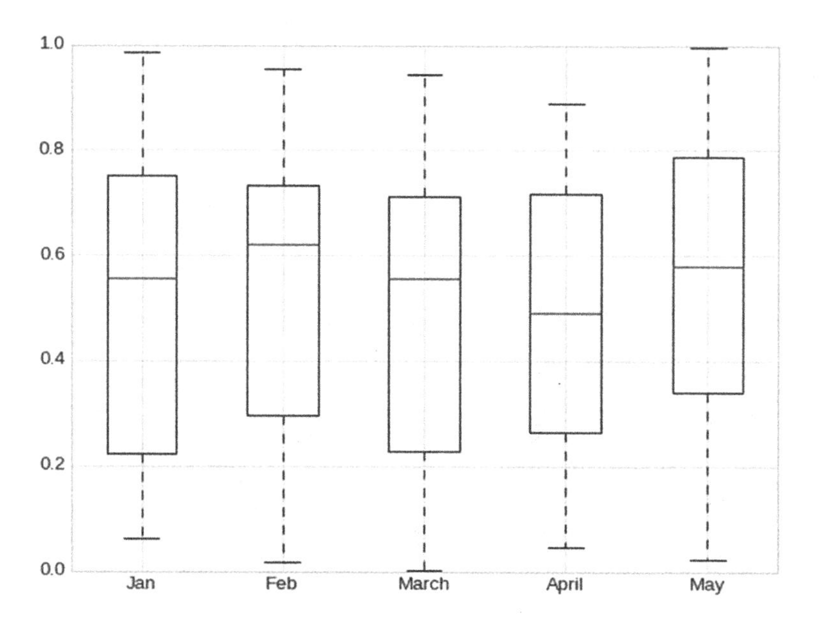

Figure 2-19. *Box plot graph*

Listing 2-14 implements an area plot (see Figure 2-20).

Listing 2-14. Creating an Area Plot

```
In [145]: import pandas as pd
          import numpy as np
          df = pd.DataFrame(np.random.rand(20,5),
               columns= ['Jan','Feb','March','April', 'May'])
          df.plot.area(figsize=(6, 4)).legend
          (bbox_to_anchor=(1.3, 1))
```

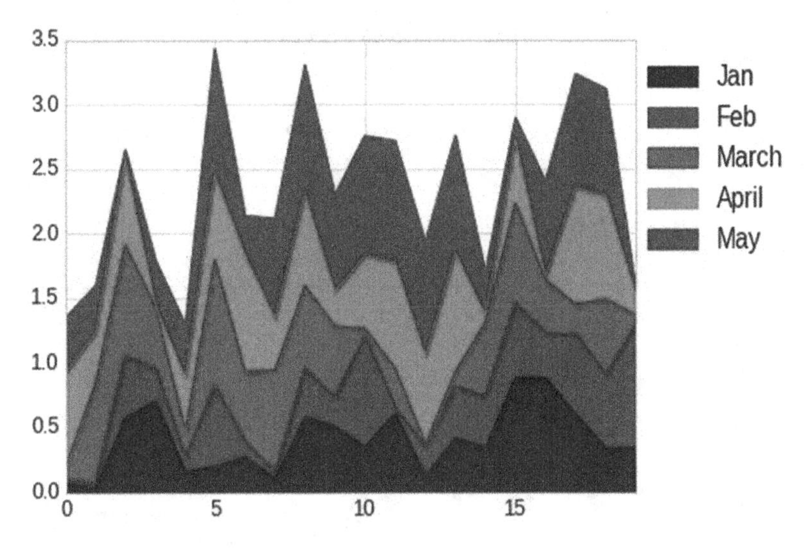

Figure 2-20. *Area plot graph*

Listing 2-15 creates a scatter plot (see Figure 2-21).

Listing 2-15. Creating a Scatter Plot

```
In [150]: import pandas as pd
import numpy as np
df = pd.DataFrame(np.random.rand(20,5),columns= ['Jan','Feb',
'March','April', 'May'])
df.plot.scatter(x='Feb', y='Jan', title='Temperature over two
months ')
```

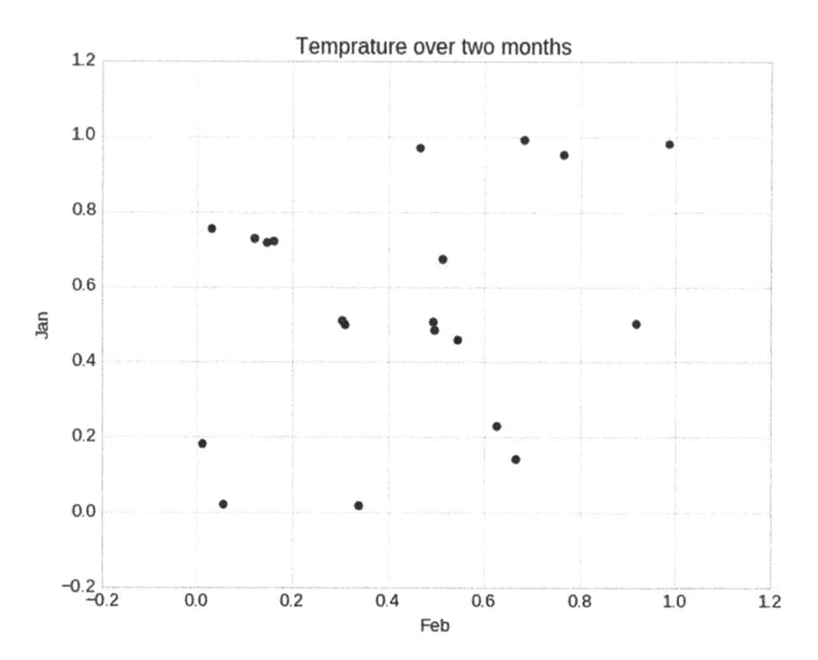

***Figure 2-21.** Scatter plot graph*

See Chapter 7 for more graphing formats.

Summary

This chapter demonstrated how to implement data visualization in modern business. Let's recap what you studied in this chapter.

- Understand the importance of data visualization.
- Acknowledge the usage of data visualization in modern business and its future implementations.
- Recognize the role of data visualization in decision-making.

- Load and use important Python data visualization libraries.
- Revise exercises with model answers for practicing and simulating real-life scenarios.

The next chapter will cover data collection structure and much more.

Exercises and Answers

1. What is meant by data visualization?

Answer:

Data visualization is the process of interpreting the data in the form of pictorial or graphical format.

2. Why is data visualization important?

Answer:

Data Visualization helps business to achieve numerous goals through the following.

- Convert the business data into interactive graphs for dynamic interpretation to serve the business goals.
- Transforming data into visually appealing, interactive dashboards of various data sources to serve the business with the insights.
- Create more attractive and informative dashboard of various graphical data representation.
- Make appropriate decisions by drilling into the data and finding the insights.
- Figure out the patterns, trends and correlations in the data being analyzed to determine where they must improve their operational processes and thereby grow their business.

- Give full picture of the data under analysis.
- Enable to organize and present massive data intuitively to present important findings from the data.
- Make better, quick and informed decisions.

3. Why do modern businesses need data visualization?

Answer:

Data visualization is needed by the modern business to support the following areas.

- Analyze the business different processes where the management can focus on the areas of improvement to generate more revenue and improve productivity.
- Bring business intelligences to life.
- Apply creative approach to improve the abilities to understand the hidden information within the business data.
- Provide better and faster way to identify patterns, trends, and correlation in the data sets that would remain undetected with a text.
- Identify new business opportunities by predicting upcoming trends or sales volumes and the revenue they would generate.
- Helps to spot trends in data that may not have been noticeable from the text alone.
- Supply managers with information they need to make more effective comparisons between data sets by plotting them on the same visualization.

- Enable managers to understand the correlations between the operating conditions and business performance.
- Help to discover the gray areas of the business and hence take right decisions for improvement.
- Helps to understand customers' behaviors and interests, and hence retains customers and market.

4. How is data visualization used for business decision-making?

Answer:

There are many ways in which visualization help the business to improve decision making.

Faster Times Response: It becomes incredibly useful to put useful interpretation of the collected data into the hands of managers and decision makers enabling them to quickly identify issues and improve response times.

Simplicity: data visualization techniques gives the full picture of the scoped parameters and simplify the data by enabling decision makers to cherry-pick the relevant data they need and dive to detailed wherever is needed.

Easier Pattern Visualization: provides easier approaches to identify upcoming trends and patterns within datasets, and hence enable to take efficient decisions and prepare strategies in advance.

Team Involvement: increase the levels of collaboration between departments and keep them on the same page to achieve strategic goals.

Unify Interpretation: produced charts and graphics have the same interpretation by all beneficial who use extracted information for decisions making and hence avoid any misleading.

5. Write a Python script to create a data frame for the following table:

Name	Mobile_Sales	TV_Sales
Ahmed	2540	2200
Omar	1370	1900
Ali	1320	2150
Ziad	2000	1850
Salwa	2100	1770
Lila	2150	2000

Answer:

```
In [ ]: import pandas as pd
       import numpy as np
       import matplotlib.pyplot as plt
salesMen = ['Ahmed', 'Omar', 'Ali', 'Ziad', 'Salwa', 'Lila']
Mobile_Sales = [2540, 1370, 1320, 2000, 2100, 2150]
TV_Sales = [2200, 1900, 2150, 1850, 1770, 2000]

df = pd.DataFrame()
df ['Name'] =salesMen
df ['Mobile_Sales'] = Mobile_Sales
df['TV_Sales']=TV_Sales

df.set_index("Name",drop=True,inplace=True)

In [13]: df
```

```
Out[13]: Name    Mobile_Sales   TV_Sales
         Ahmed   2540           2200
         Omar    1370           1900
         Ali     1320           2150
         Ziad    2000           1850
         Salwa   2100           1770
         Lila    2150           2000
```

For the created data frame in the previous question, do the following:

A. Create a bar plot of the sales volume.

Answer:

```
In [5]: df.plot.bar( figsize=(20, 10), rot=0).legend(bbox_to_
anchor=(1.1, 1)) plt.xlabel('Salesmen') plt.ylabel('Sales')
plt.title('Sales Volume for two salesmen in \nJanuary and April 2017')
plt.show()
```

See also Figure 2-22.

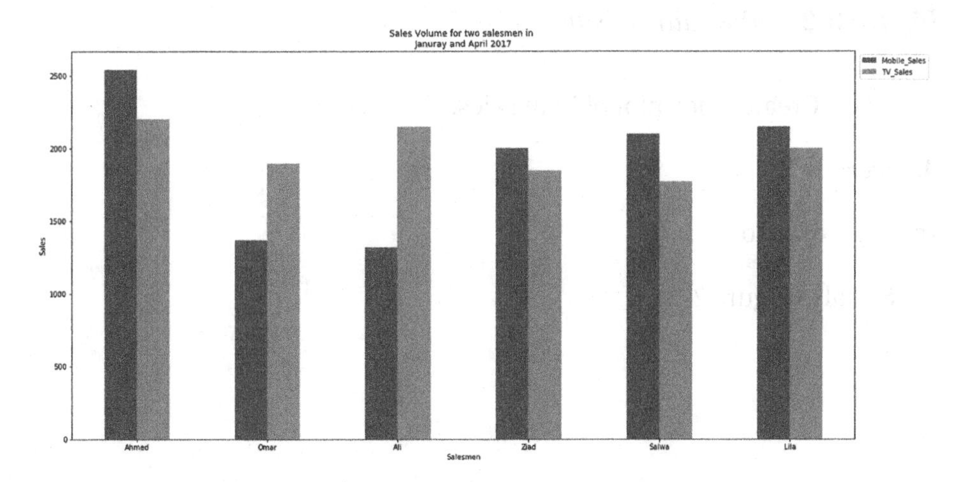

Figure 2-22. *Bar plot of sales*

B. Create a pie chart of item sales.

Answer:

```
In [6]: df.plot.pie(subplots=True)
```

See also Figure 2-23.

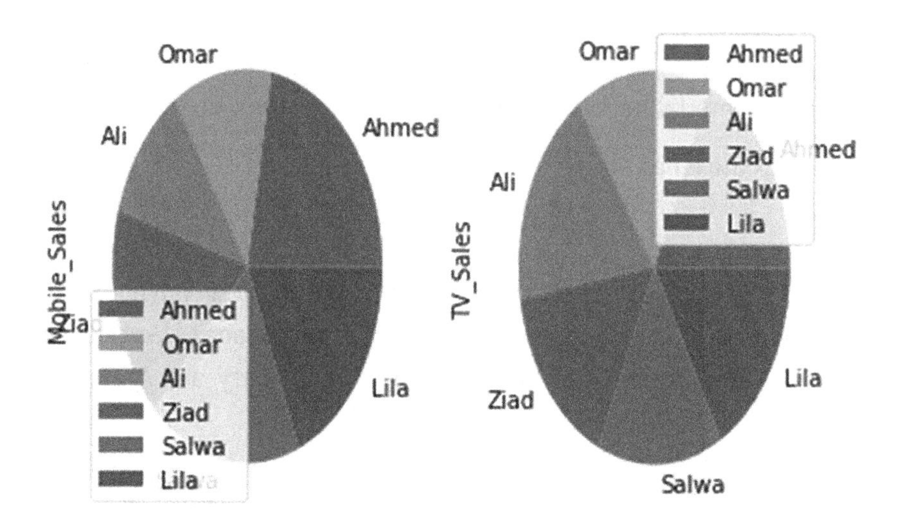

Figure 2-23. *Pie chart of sales*

C. Create a box plot of item sales.

Answer:

```
In [8]: df.plot.box()
```

See also Figure 2-24.

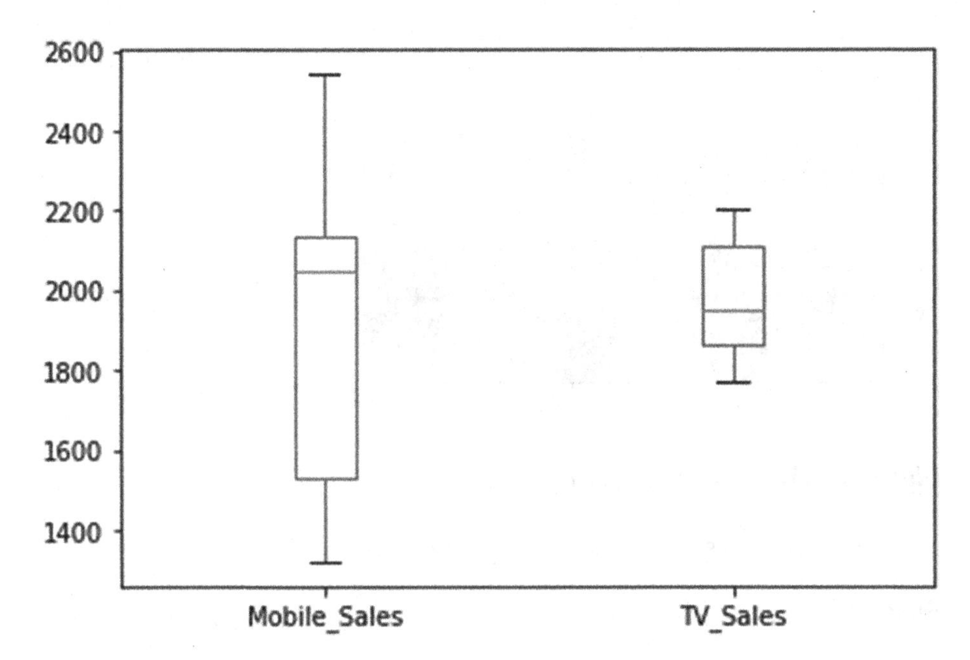

Figure 2-24. *Box plot of sales*

D. Create an area plot of item sales.

Answer:

```
In [9]: df.plot.area(figsize=(6, 4)).legend(bbox_to_anchor=(1.3,
                    1))
```

See also Figure 2-25.

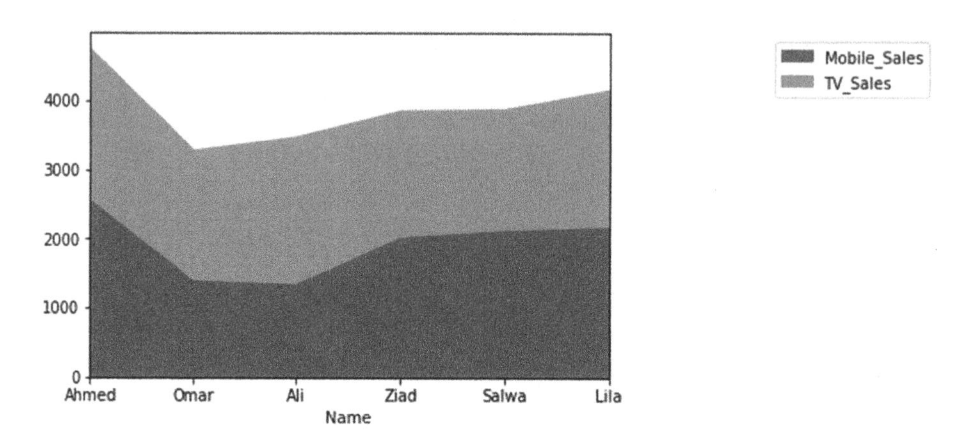

Figure 2-25. *Area plot of sales*

E. Create a stacked bar plot of item sales.

Answer:

```
In [11]: df.plot.bar(stacked=True, figsize=(20, 10)).legend
                      (bbox_to_anchor=(1.1, 1))
```

See also Figure 2-26.

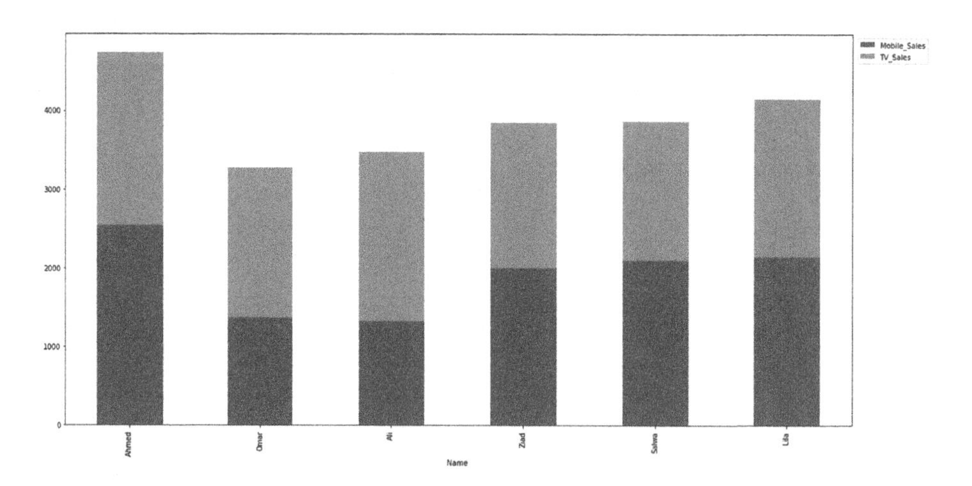

Figure 2-26. *Stacked bar plot of sales*

CHAPTER 3

Data Collection Structures

Lists, dictionaries, tuples, series, data frames, and panels are Python data collection structures that can be used to maintain a collection of data. This chapter will demonstrate these various structures in detail with practical examples.

Lists

A *list* is a sequence of values of any data type that can be accessed forward or backward. Each value is called an *element* or a *list item*. Lists are mutable, which means that you won't create a new list when you modify a list element. Elements are stored in the given order. Various operations can be conducted on lists such as insertion, sort, and deletion. A list can be created by storing a sequence of different types of values separated by commas. A Python list is enclosed between a square brackets ([ ]), and elements are stored in the index based on a starting index of 0.

O. Embarak, *Data Analysis and Visualization Using Python*,
https://doi.org/10.1007/978-1-4842-4109-7_3

Creating Lists

You can have lists of string values and integers, empty lists, and nested lists, which are lists inside other lists. Listing 3-1 shows how to create a list.

Listing 3-1. Creating Lists

```
In [1]: # Create List
        List1 = [1, 24, 76]
        print (List1)
        colors=['red', 'yellow', 'blue']
        print (colors)
        mix=['red', 24, 98.6]
        print (mix)
        nested= [ 1, [5, 6], 7]
        print (nested)
        print ([])

[1, 24, 76]
['red', 'yellow', 'blue']
['red', 24, 98.6]
[1, [5, 6], 7]
[]
```

Accessing Values in Lists

You can access list elements forward or backward. For instance, in Listing 3-2, `list2 [3: ]` returns elements starting from index 3 to the end of the list since `list2` has four elements where `[4,5]` is the element of index 3, which is in the form of nested list. Then you get `[[4,5]]`

as a result of `print (list2 [3:])`. You can also access a list element backward using negative indices. For example, `list3[-3]` will return the third element in the backward sequence `n-3`, i.e., index 1. Here's an example:

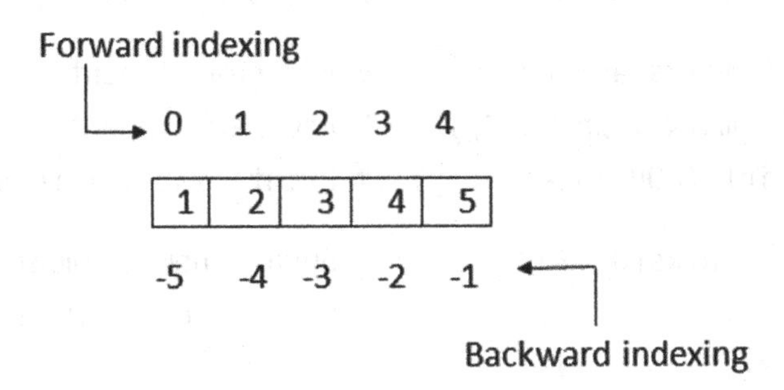

Listing 3-2. Accessing Lists

```
In [9]: list1 = ['Egypt', 'chemistry', 2017, 2018]
        list2 = [1, 2, 3, [4, 5] ]
        list3 = ["a", 3.7, '330', "Omar"]
        print (list1[2])
        print (list2 [3:])
        print (list3 [-3:-1])
        print (list3[-3])

        2017
        [[4, 5]]
        [3.7, '330']
        3.7
```

Adding and Updating Lists

You can update single or multiple elements of lists by giving the slice on the left side of the assign operator, and you can add elements to a list with the `append()` method, as shown in Listing 3-3.

Listing 3-3. Adding and Updating List Elements

```
In [50]: courses=["OOP","Networking","MIS","Project"]
students=["Ahmed", "Ali",
         "Salim", "Abdullah", "Salwa"] OOP_marks = [65, 85, 92]

         OOP_marks.append(50)    # Add new element
         OOP_marks.append(77)    # Add new element
         print (OOP_marks[ : ])  # Print list before updating

         OOP_marks[0]=70         # update new element
         OOP_marks[1]=45         # update new element
         list1 = [88, 93]
         OOP_marks.extend(list1) # extend list with another
         list print
         (OOP_marks[ : ])        # Print list after updating
[65, 85, 92, 50, 77]
[70, 45, 92, 50, 77, 88, 93]
```

As shown in Listing 3-3, you can add a new element to the list using the `append()` method. You can also update an element in the list by using the list name and the element index. For example, `OOP_marks[1]=45` changes the value of index 1 from 85 to 45.

Deleting List Elements

To remove a list element, either you can delete it using the `del` statement in the element index, or you can remove the element using the `remove()` method via the element value in the list. If you use the `remove()` method to remove an element that is repeated more than one time in the list, it removes only the first occurrence of that element inside the list. Also, you can use the `pop()` method to remove a specific element by its index value, as shown in Listing 3-4.

Listing 3-4. Deleting an Element from a List

```
In [48]: OOP_marks = [70, 45, 92, 50, 77, 45]
         print (OOP_marks)

         del OOP_marks[0] # delete an element using del
         print (OOP_marks)

         OOP_marks.remove (45) # remove an element using
         remove() method
         print (OOP_marks)

         OOP_marks.pop (2) # remove an element using pop()
         method
         print (OOP_marks)

         [70, 45, 92, 50, 77, 45]
         [45, 92, 50, 77, 45]
         [92, 50, 77, 45]
         [92, 50, 45]
```

Basic List Operations

Like string processing, lists respond to + and * operators as concatenation and repetition, except that the result is a new list, as shown in Listing 3-5.

Listing 3-5. List Operations

```
In [46]:print (len([5, "Omar", 3]))      # find the list
length.
         print ([3, 4, 1] + ["Omar", 5, 6]) # concatenate lists.
         print (['Eg!'] * 4)  # repeat an element in a list.
         print (3 in [1, 2, 3])     # check if element in a list
         for x in [1, 2, 3]:
              print (x, end=' ') # traverse list elements
```

```
3
[3, 4, 1, 'Omar', 5, 6]
['Eg!', 'Eg!', 'Eg!', 'Eg!']
True
1 2 3
```

Indexing, Slicing, and Matrices

Lists are a sequence of indexed elements that can be accessed forward or backward. Therefore, you can read their elements using a positive index or negative (backward) index, as shown in Listing 3-6.

Listing 3-6. Indexing and Slicing List Elements

```
In [9]:list1 = ['Egypt', 'chemistry', 2017, 2018]
       list2 = [1, 2, 3, [4, 5]]
       list3 = ["a", 3.7, '330', "Omar"]

       print (list1[2])
       print (list2 [3:])
       print (list3 [-3:-1])
       print (list3[-3])
       2017
       [[4, 5]]
       [3.7, '330']
       3.7
```

Built-in List Functions and Methods

Various functions and methods can be used for list processing, as shown in Table 3-1.

Table 3-1. *List Functions*

Sr.No.	Function	Description
1	`cmp(list1, list2)`	Compares elements of both lists
2	`len(list1)`	Gives the total length of the list
3	`max(list1)`	Returns an item from the list with `max` value
4	`min(list1)`	Returns an item from the list with `min` value
5	`list(seq)`	Converts a tuple into list

List Functions

Built-in functions facilitate list processing. The following tables show functions and methods that can be used to manipulate lists. For example, you can simply use `cmp()` to compare two lists, and if both are identical, it returns `TRUE`; otherwise, it returns `FALSE`. You can find the list size using the `len()` method. In addition, you can find the minimum and maximum values in a list using the `min()` and `max()` methods, respectively. See Listing 3-7 for an example.

Listing 3-7. A Python Script to Apply List Functions

```
In [51]: #Built-in Functions and Lists
tickets = [3, 41, 12, 9, 74, 15]
         print (tickets)
         print (len(tickets))
         print (max(tickets))
         print (min(tickets))
         print (sum(tickets))
         print (sum(tickets)/len(tickets))

         [3, 41, 12, 9, 74, 15]
         6
```

```
74
3
154
25.666666666666668
```

List Methods

Built-in methods facilitate list editing. Table 3-2 shows that you can simply use append(), insert(), and extend() to add new elements to the list. The pop() and remove() methods are used to remove elements from a list. Table 3-2 summarizes some methods that you can adapt to the created list.

Table 3-2. *Built-in List Methods*

Sr.No.	Methods	Description
1	list.append(obj)	Appends object obj to the list
2	list.count(obj)	Returns count of how many times obj occurs in the list
3	list.extend(seq)	Appends the contents of seq to the list
4	list.index(obj)	Returns the lowest index in the list that obj appears in
5	list.insert(index, obj)	Inserts object obj into the list at offset index
6	list.pop(obj=list[-1])	Removes and returns last object or obj from list
7	list.remove(obj)	Removes object obj from list
8	list.reverse()	Reverses objects of list in place
9	list.sort([func])	Sorts objects of list; use compare func if given

List Sorting and Traversing

Sorting lists is important, especially for list-searching purposes. You can create a list from a sequence; in addition, you can sort and traverse list elements for processing using iteration statements, as shown in Listing 3-8.

Listing 3-8. List Sorting and Traversing

```
In [58]: #List sorting and Traversing
         seq=(41, 12, 9, 74, 3, 15) # use sequence for creating
         a list
         tickets=list(seq)

         print (tickets)
         tickets.sort()
         print (tickets)

         print ("\nSorted list elements ")
         for ticket in tickets:
              print (ticket)
        [41, 12, 9, 74, 3, 15]
        [3, 9, 12, 15, 41, 74]

        Sorted list elements
        3
        9
        12
        15
        41
        74
```

Lists and Strings

You can split a string into a list of characters. In addition, you can split a string into a list of words using the `split()` method. The default delimiter for the `split()` method is a white space. However, you can specify which characters to use as the word boundaries. For example, you can use a hyphen as a delimiter, as in Listing 3-9.

Listing 3-9. Converting a String into a List of Characters or Words

```
In [63]: # convert string to a list of characters
         Word = 'Egypt'
         List1 = list(Word)
         print (List1)
         ['E', 'g', 'y', 'p', 't']

In [69]: # use the delimiter
         Greeting= 'Welcome-to-Egypt'
         List2 =Greeting.split("-")
         print (List2)

         Greeting= 'Welcome-to-Egypt'
         delimiter='-'
         List2 =Greeting.split(delimiter)
         print (List2)
         ['Welcome', 'to', 'Egypt']
         ['Welcome', 'to', 'Egypt']

In [70]: # we can break a string into words using the split
method
         Greeting= 'Welcome to Egypt'
         List2 =Greeting.split()
         print (List2)
```

```
print (List2[2])
['Welcome', 'to', 'Egypt']
Egypt
```

The `join()` method is the inverse of the split method (see Listing 3-10). It takes a list of strings and concatenates the elements. You have to specify the delimiter that the `join()` method will add between the list elements to form a string.

Listing 3-10. Using the join() Method

```
In [73]: List1 = ['Welcome', 'to', 'Egypt']
         delimiter = ' '
         delimiter.join(List1)
Out[73]: 'Welcome to Egypt'

In [74]: List1 = ['Welcome', 'to', 'Egypt']
         delimiter = '-'
         delimiter.join(List1)
Out[74]: 'Welcome-to-Egypt'
```

Parsing Lines

You can read text data from a file and convert it into a list of words for further processing. Figure 3-1 shows that you can read `myfile.txt`, parse it line per line, and convert the data into a list of words.

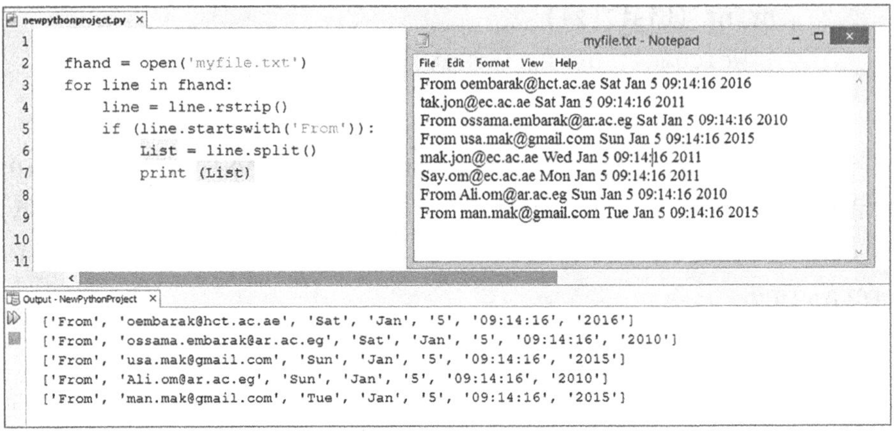
newpythonproject.py
fhand = open('myfile.txt')
for line in fhand:
line = line.rstrip()
if (line.startswith('From')):
List = line.split()
print (List)
myfile.txt - Notepad
File Edit Format View Help
From oembarak@hct.ac.ae Sat Jan 5 09:14:16 2016
tak.jon@ec.ac.ae Sat Jan 5 09:14:16 2011
From ossama.embarak@ar.ac.eg Sat Jan 5 09:14:16 2010
From usa.mak@gmail.com Sun Jan 5 09:14:16 2015
mak.jon@ec.ac.ae Wed Jan 5 09:14:16 2011
Say.om@ec.ac.ae Mon Jan 5 09:14:16 2011
From Ali.om@ar.ac.eg Sun Jan 5 09:14:16 2010
From man.mak@gmail.com Tue Jan 5 09:14:16 2015
Output - NewPythonProject
['From', 'oembarak@hct.ac.ae', 'Sat', 'Jan', '5', '09:14:16', '2016']
['From', 'ossama.embarak@ar.ac.eg', 'Sat', 'Jan', '5', '09:14:16', '2010']
['From', 'usa.mak@gmail.com', 'Sun', 'Jan', '5', '09:14:16', '2015']
['From', 'Ali.om@ar.ac.eg', 'Sun', 'Jan', '5', '09:14:16', '2010']
['From', 'man.mak@gmail.com', 'Tue', 'Jan', '5', '09:14:16', '2015']

Figure 3-1. *Parsing text lines*

In the previous example, you can extract only years or e-mails of contacts, as shown in Figure 3-2.

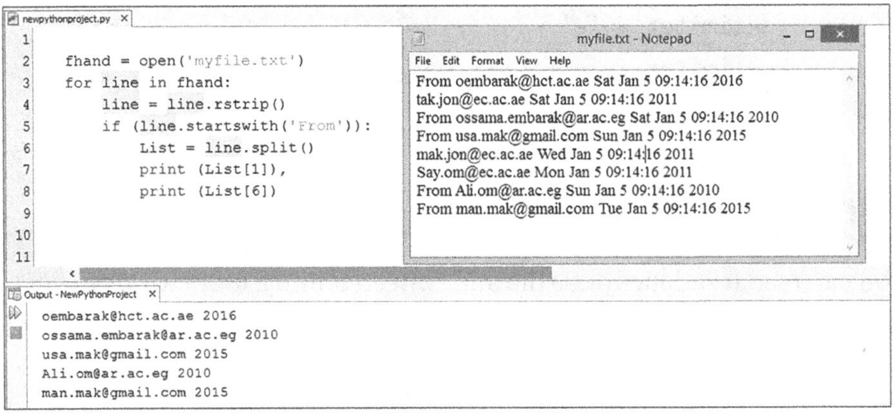
newpythonproject.py
fhand = open('myfile.txt')
for line in fhand:
line = line.rstrip()
if (line.startswith('From')):
List = line.split()
print (List[1]),
print (List[6])
myfile.txt - Notepad
File Edit Format View Help
From oembarak@hct.ac.ae Sat Jan 5 09:14:16 2016
tak.jon@ec.ac.ae Sat Jan 5 09:14:16 2011
From ossama.embarak@ar.ac.eg Sat Jan 5 09:14:16 2010
From usa.mak@gmail.com Sun Jan 5 09:14:16 2015
mak.jon@ec.ac.ae Wed Jan 5 09:14:16 2011
Say.om@ec.ac.ae Mon Jan 5 09:14:16 2011
From Ali.om@ar.ac.eg Sun Jan 5 09:14:16 2010
From man.mak@gmail.com Tue Jan 5 09:14:16 2015
Output - NewPythonProject
oembarak@hct.ac.ae 2016
ossama.embarak@ar.ac.eg 2010
usa.mak@gmail.com 2015
Ali.om@ar.ac.eg 2010
man.mak@gmail.com 2015

Figure 3-2. *Extracting specific data from a text file via lists*

Aliasing

The assign operator is dangerous if you don't use it carefully. The association of a variable with an object is called a *reference*. In addition, an object with more than one reference and more than one name is called an *alias*. Listing 3-11 demonstrates the use of the assign operator. Say you have a list called a. If a refers to an object and you assign b = a, then both variables a and b refer to the same object, and an operation conducted on a will automatically adapt to b.

Listing 3-11. *Alias Objects*

With Alias	Without Alias
In [117]:a = [1, 2, 3] b = a print (a) print (b)	In [120]:a = [1, 2, 3] b = [1, 2, 3] print (a) print (b)
[1, 2, 3] [1, 2, 3]	[1, 2, 3] [1, 2, 3]
In [118]:a.append(77) print (a) print (b)	In [121]:a.append(77) print (a) print (b)
[1, 2, 3, 77] [1, 2, 3, 77]	[1, 2, 3, 77] [1, 2, 3]
In [119]: b is a	In [122]: b is a
Out[119]: True	Out[122]: False

Dictionaries

A *dictionary* is an unordered set of key-value pair; each key is separated from its value by a colon (:). The items (the pair) are separated by commas, and the whole thing is enclosed in curly braces ({}). In fact, an empty dictionary is written only with curly braces:. Dictionary keys should be unique and should be of an immutable data type such as string, integer, etc.

Dictionary values can be repeated many times, and the values can be of any data type. It's a mapping between keys and values; you can create a dictionary using the dict() method.

Creating Dictionaries

You can create a dictionary and assign a key-value pair directly. In addition, you can create an empty dictionary and then assign values to each generated key, as shown in Listing 3-12.

Listing 3-12. Creating Dictionaries

```
In [36]: Prices = {"Honda":40000, "Suzuki":50000,
"Mercedes":85000, "Nissan":35000, "Mitsubishi": 43000}
         print (Prices)
         {'Honda': 40000, 'Suzuki': 50000, 'Mercedes': 85000,
         'Nissan': 35000, 'Mitsubishi': 43000}

In [37]: Staff_Salary = { 'Omar Ahmed' : 30000 , 'Ali Ziad' :
                          24000,
                        'Ossama Hashim': 25000,
                        'Majid Hatem':10000}
         print(Staff_Salary)
         STDMarks={"Salwa Ahmed":50, "Abdullah Mohamed":80,
         "Sultan Ghanim":90}
         print(STDMarks)
         {'Omar Ahmed': 30000, 'Ali Ziad': 24000,
         'Ossama Hashim': 25000, 'Majid Hatem': 10000}
         {'Salwa Ahmed': 50, 'Abdullah Mohamed': 80,
         'Sultan Ghanim': 90}
```

```
In [38]:STDMarks = dict()
        STDMarks['Salwa Ahmed']=50
        STDMarks['Abdullah Mohamed']=80
        STDMarks['Sultan Ghanim']=90
        print (STDMarks)
        {'Salwa Ahmed': 50, 'Abdullah Mohamed': 80, 'Sultan
        Ghanim': 90}
```

Updating and Accessing Values in Dictionaries

Once you have created a dictionary, you can update and access its values for any further processing. Listing 3-13 shows that you can add a new item called `STDMarks['Omar Majid'] = 74` where `Omar Majid` is the key and 74 is the value mapped to that key. Also, you can update the existing value of the key `Salwa Ahmed`.

Listing 3-13. Updating and Adding a New Item to a Dictionary

```
In [39]: STDMarks={"Salwa Ahmed":50, "Abdullah Mohamed":80,
                   "Sultan
                    Ghanim":90}
        STDMarks['Salwa Ahmed'] = 85 # update current value of
        the key 'Salwa Ahmed'
        STDMarks['Omar Majid'] = 74 # Add a new item to the
        dictionary
        print (STDMarks)
        {'Salwa Ahmed': 85, 'Abdullah Mohamed': 80, 'Sultan
        Ghanim': 90, 'Omar Majid': 74}
```

You can directly access any element in the dictionary or iterate all dictionary elements, as shown in Listing 3-14.

Listing 3-14. Accessing Dictionary Elements

```
In [2]: Staff_Salary = { 'Omar Ahmed' : 30000 , 'Ali Ziad' :
24000, 'Ossama Hashim': 25000, 'Majid Hatem':10000}
        print('Salary package for Ossama Hashim is ', end=“)

        # access specific dictionary element
        print(Staff_Salary['Ossama Hashim'])

        Salary package for Ossama Hashim is 25000


In [3]: # Define a function to return salary after discount tax
5% def Netsalary (salary):
            return salary - (salary * 0.05) # also, could be
            return salary *0.95
        #Iterate all elements in a dictionary
        print ("Name" , '\t', "Net Salary" )
        for key, value in Staff_Salary.items():
               print (key , '\t', Netsalary(value))
         Name              Net Salary
         Omar Ahmed     28500.0
         Ali Ziad       22800.0
         Ossama Hashim 23750.0
         Majid Hatem    9500.0
```

Listing 3-14 shows that you can create a function to calculate the net salary after deducting the salary tax value of 5 percent, and you iterate all dictionary elements. In each iteration, you print the key name and the returned net salary value.

Deleting Dictionary Elements

You can either remove individual dictionary elements using the element key or clear the entire contents of a dictionary. Also, you can delete the entire dictionary in a single operation using a `del` keyword, as shown in Listing 3-15. It should be noted that it's not allowed to have repeated keys in a dictionary.

Listing 3-15. Alter a Dictionary

```
In [40]: STDMarks={"Salwa Ahmed":50, "Abdullah Mohamed":80,
"Sultan Ghanim":90}
         print (STDMarks)
         del STDMarks['Abdullah Mohamed'] # remove entry with
         key 'Abdullah Mohamed'
         print (STDMarks)
         STDMarks.clear() # remove all entries in STDMarks
         dictionary
         print (STDMarks)
         del STDMarks  # delete entire dictionary
         {'Salwa Ahmed': 50, 'Abdullah Mohamed': 80, 'Sultan
         Ghanim': 90}
         {'Salwa Ahmed': 50, 'Sultan Ghanim': 90}
         {}
```

Built-in Dictionary Functions

Various built-in functions can be implemented on dictionaries. Table 3-3 shows some of these functions. The compare function `cmp()` in older Python versions was used to compare two dictionaries; it returns 0 if both dictionaries are equal, 1 if `dic1` > `dict2`, and -1 if `dict1` < `dict2`. But starting with Python 3, the `cmp()` function is not available anymore, and you cannot define it. See also Listing 3-16.

Table 3-3. *Built-in Dictionary Functions*

No	Function	Description
1	cmp(dict1, dict2)	Compares elements of two dictionaries.
2	len(dict)	Gives the total length of the dictionary, i.e., the number of items in the dictionary.
3	str(dict)	Produces a printable string representation of a dictionary.
4	type(variable)	Returns the type of the passed variable. If the passed variable is a dictionary, then it would return a dictionary type.

Listing 3-16. Implementing Dictionary Functions

```
In [43]:Staff_Salary = { 'Omar Ahmed' : 30000 , 'Ali Ziad' :
                        24000,
                       'Ossama Hashim': 25000, 'Majid
                       Hatem':10000}
        STDMarks={"Salwa Ahmed":50, "Abdullah Mohamed":80,
                  "Sultan
                   Ghanim":90}

In [52]: def cmp(a, b):
               for key, value in a.items():
                   for key1, value1 in b.items():
                        return (key >key1) - (key < key1)

In [54]: print (cmp(STDMarks,Staff_Salary) )
          print (cmp(STDMarks,STDMarks) )
          print (len(STDMarks) )
          print (str(STDMarks) )
          print (type(STDMarks) )
          1
```

```
0
3
{'Salwa Ahmed': 50, 'Abdullah Mohamed': 80, 'Sultan
Ghanim': 90}
<class 'dict'>
```

Built-in Dictionary Methods

Python provides various methods for dictionary processing. Table 3-4 summarizes the methods that can be used to access dictionaries.

Table 3-4. *Built-in Dictionary Methods*

No	Methods	Description
1	dict1.clear()	Removes all elements of dictionary dict1
2	dict1.copy()	Returns a copy of dictionary dict1
3	dict1.fromkeys()	Creates a new dictionary with keys from seq and values
4	dict1.get(key, default=None)	For the key name key, returns the value or default if key not in dictionary
5	dict1.has_key(key)	Returns true if key is in dictionary dict1, false otherwise
6	dict1.items()	Returns a list of dict1's (key, value) tuple pairs
7	dict1.keys()	Returns list of the dictionary dict1's keys
8	dict1.setdefault(key, default=None)	Similar to get(), but will set dict1[key]=default if key is not already in dict1
9	dict1.update(dict2)	Adds dictionary dict2's key-values pairs to dict1
10	dict1.values()	Returns list of dictionary dict1's values

Listing 3-17 shows the use and implementation of dictionary methods.

Listing 3-17. Implementing Dictionary Methods

```
In [89]: Staff_Salary = { 'Omar Ahmed' : 30000 , 'Ali Ziad' :
                          24000,
                         'Ossama Hashim': 25000, 'Majid
                         Hatem':10000}
         STDMarks={"Salwa Ahmed":50, "Abdullah Mohamed":80,
                   "Sultan
                    Ghanim":90}
         print (Staff_Salary.get('Ali Ziad') )
         print (STDMarks.items())
         print (Staff_Salary.keys())

         print()
         STDMarks.setdefault('Ali Ziad')
         print (STDMarks)
         print (STDMarks.update(dict1))
         print (STDMarks)
         24000
         dict_items([('Salwa Ahmed', 50), ('Abdullah Mohamed',
         80), ('Sultan Ghanim', 90)])
         dict_keys(['Omar Ahmed', 'Ali Ziad', 'Ossama Hashim',
         'Majid Hatem'])
         {'Salwa Ahmed': 50, 'Abdullah Mohamed': 80, 'Sultan
         Ghanim': 90, 'Ali Ziad': None}
         None
         {'Salwa Ahmed': 50, 'Abdullah Mohamed': 80, 'Sultan
         Ghanim': 90, 'Ali Ziad': None}
```

You can sort a dictionary by key and by value, as shown in Listing 3-18.

Listing 3-18. Sorting a Dictionary

```
In [96]: Staff_Salary = { 'Omar Ahmed' : 30000 , 'Ali Ziad' :
24000, 'Ossama Hashim': 25000, 'Majid Hatem':10000}
       print ("\nSorted by key")
       for k in sorted(Staff_Salary):
           print (k, Staff_Salary[k])
       Sorted by key
       Ali Ziad 24000
       Majid Hatem 10000
       Omar Ahmed 30000
       Ossama Hashim 25000
```

```
In [97]: Staff_Salary = { 'Omar Ahmed' : 30000 , 'Ali Ziad' :
24000, 'Ossama Hashim': 25000, 'Majid Hatem':10000}
       print ("\nSorted by value")
       for w in sorted(Staff_Salary, key=Staff_Salary.get,
       reverse=True):
               print (w, Staff_Salary[w])
       Sorted by value
       Omar Ahmed      30000
       Ossama Hashim   25000
       Ali Ziad        24000
       Majid Hatem     10000
```

Tuples

A *tuple* is a sequence just like a list of immutable objects. The differences between tuples and lists are that the tuples cannot be altered; also, tuples use parentheses, whereas lists use square brackets.

Creating Tuples

You can create tuples simply by using different comma-separated values. You can access an element in the tuple by index, as shown in Listing 3-19.

Listing 3-19. Creating and Displaying Tuples

```
In [1]:Names = ('Omar', 'Ali', 'Bahaa')
       Marks = ( 75, 65, 95 )
       print (Names[2])
       print (Marks)
       print (max(Marks))
       Bahaa
       (75, 65, 95)
       95

In [2]: for name in Names:
              print (name)
         Omar
         Ali
         Bahaa
```

Let's try to alter a tuple to modify any element, as shown in Listing 3-20; we get an error because, as indicated earlier, tuples cannot be altered.

Listing 3-20. Altering a Tuple for Editing

```
In [3]: Marks[1]=66
        -------------------------------------------------------
        TypeError    Traceback (most recent call last)
        <ipython-input-3-b225998b9edb> in <module>()
        ----> 1 Marks[1]=66
        TypeError: 'tuple' object does not support item
        assignment
```

Like lists, you can access tuple elements forward and backward using the element's indices. Here's an example:

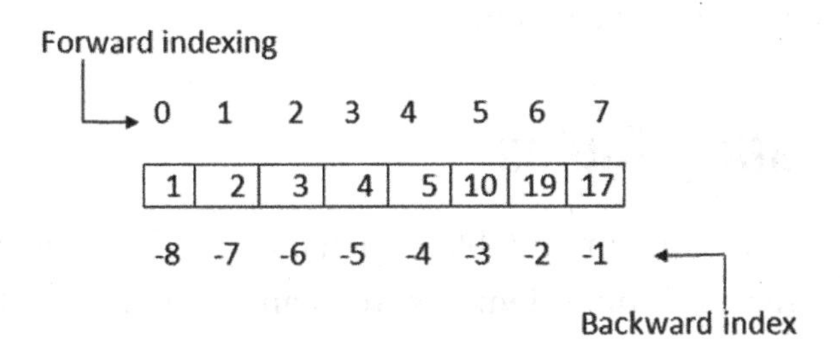

You can sort a list of tuples. Listing 3-21 shows how to sort tuple elements in place as well as how to create another sorted tuple.

Listing 3-21. Sorting a Tuple

```
In [1]:import operator
      MarksCIS = [(88,65),(70,90,85), (55,88,44)]
      print (MarksCIS)            # original tuples
      print (sorted(MarksCIS))    # direct sorting
      [(88, 65), (70, 90, 85), (55, 88, 44)]
      [(55, 88, 44), (70, 90, 85), (88, 65)]

In [2]: print (MarksCIS)      # original tuples
        #create a new sorted tuple
        MarksCIS2 = sorted(MarksCIS, key=lambda x: (x[0], x[1]))
        print (MarksCIS2)
        [(88, 65), (70, 90, 85), (55, 88, 44)]
        [(55, 88, 44), (70, 90, 85), (88, 65)]
In [3]:print (MarksCIS) # original tuples
       MarksCIS.sort(key=lambda x: (x[0], x[1])) # sort in tuple
        print (MarksCIS)
        [(88, 65), (70, 90, 85), (55, 88, 44)]
        [(55, 88, 44), (70, 90, 85), (88, 65)]
```

By default the `sort` built-in function detected that the items are in tuples form, so the sort function sorts tuples based on the first element, then based on the second element.

Concatenating Tuples

As mentioned, tuples are immutable, which means you cannot update or change the values of tuple elements. You can take portions of existing tuples to create new tuples, as Listing 3-22 demonstrates.

Listing 3-22. Concatenating Tuples

```
In [5]:MarksCIS=(70,85,55)
        MarksCIN=(90,75,60)
        Combind=MarksCIS + MarksCIN
        print (Combind)
        (70, 85, 55, 90, 75, 60)
```

Accessing Values in Tuples

To access an element in a tuple, you can use square brackets and the element index for retrieving an element value, as shown in Listing 3-23.

Listing 3-23. Accessing Values in a Tuple

```
In [4]:MarksCIS = (70, 85, 55)
        MarksCIN = (90, 75, 60)
       print ("The third mark in CIS is ", MarksCIS[2])
       print ("The third mark in CIN is ", MarksCIN[2])
       The third mark in CIS is 55
       The third mark in CIN is 60
```

You can delete a tuple using de, as shown in Listing 3-24.

Listing 3-24. Deleting a Tuple

```
In [5]: MarksCIN = (90, 75, 60)
         print (MarksCIN)
         del MarksCIN
         print (MarksCIN)
         (90, 75, 60)
         -------------------------------------------------------
         NameError                                   Traceback
                                                     (most recent
                                                     call last)
         <ipython-input-5-4c08fec39768> in <module>()
          2 print (MarksCIN) 3 del MarksCIN
    ----> 4 print (MarksCIN)
        NameError: name 'MarksCIN' is not defined
```

You received an error because you ordered Python to print a tuple named MarksCIN, which has been removed. You can access a tuple element forward and backward; in addition, you can slice values from a tuple using indices. Listing 3-25 shows that you can slice in a forward manner where MarksCIS[1:4] retrieves elements from element 1 up to element 3, while MarksCIS[:] retrieves all elements in a tuple. In backward slicing, MarksCIS[-3] retrieves the third element backward, and MarksCIS[-4:-2] retrieves the fourth element backward up to the third element but not the second backward element.

Listing 3-25. Slicing Tuple Values

```
In [6]: MarksCIS = (88, 65, 70,90,85,45,78,95,55)
        print ("\nForward slicing")
        print (MarksCIS[1:4])
        print (MarksCIS[:3])
        print (MarksCIS[6:])
        print (MarksCIS[4:6])
```

```
print ("\nBackward slicing")
print (MarksCIS[-4:-2])
print (MarksCIS[-3])
print (MarksCIS[-3:])
print (MarksCIS[ :-3])
Forward slicing
(65, 70, 90)
(88, 65, 70)
(78, 95, 55)
(85, 45)
Backward slicing
(45, 78)
78
(78, 95, 55)
(88, 65, 70, 90, 85, 45)
```

Basic Tuples Operations

Like strings, tuples respond to the + and * operators as concatenation and repetition to get a new tuple. See Table 3-5.

Table 3-5. *Tuple Operations*

Expression	Results	Description
`len((5, 7, 2,6))`	`4`	Length
`(1, 2, 3,10) + (4, 5, 6,7)`	`(1, 2, 3,10, 4, 5, 6,7)`	Concatenation
`('Hi!',) * 4`	`('Hi!', 'Hi!', 'Hi!', 'Hi!')`	Repetition
`10 in (10, 2, 3)`	`True`	Membership
`for x in (10, 1, 5): print x,`	`10 1 5`	Iteration

Series

A series is defined as a one-dimensional labeled array capable of holding any data type (integers, strings, floating-point numbers, Python objects, etc.).

```
SeriesX = pd.Series(data, index=index),
```

Here, `pd` is a Pandas form, and `data` refers to a Python dictionary, an ndarray, or even a scalar value.

Creating a Series with index

If the data is an ndarray, then the index is a list of axis labels that is directly passed; otherwise, an auto index is created by Python starting with 0 up to n-1. See Listing 3-26 and Listing 3-27.

Listing 3-26. Creating a Series of Ndarray Data with Labels

```
In [8]: import numpy as np
        import pandas as pd
        Series1 = pd.Series(np.random.randn(4), index=['a',
        'b', 'c', 'd'])
        print(Series1)
        print(Series1.index)
        a 0.350241
        b -1.214802
        c 0.704124
        d 0.866934
        dtype: float64
        Index(['a', 'b', 'c', 'd'], dtype='object')
```

Listing 3-27. Creating a Series of Ndarray Data Without Labels

```
In [9]:import numpy as np
        import pandas as pd
        Series2 = pd.Series(np.random.randn(4))
        print(Series2)
        print(Series2.index)
        0 1.784219
        1 -0.627832
        2 0.429453
        3 -0.473971
        dtype: float64
        RangeIndex(start=0, stop=4, step=1)
```

Creating a series from ndarrays is valid to most Numpy functions; also, operations such as slicing will slice the index. See Listing 3-28 and Listing 3-29.

Listing 3-28. Slicing Data from a Series

```
In [10]: print (" \n Series slicing ")
         print (Series1[:3])
         print ("\nIndex accessing")
         print (Series1[[3,1,0]])
         print ("\nSingle index")
         x = Series1[0]
         print (x)
          Series slicing
          a 0.350241
          b -1.214802
          c 0.704124
         dtype: float64
```

```
Index accessing
d 0.866934
b -1.214802
a 0.350241
dtype: float64

Single index
0.35024081401881596
```

Listing 3-29. Sample Operations in a Series

```
In [11]: print ("\nSeries Sample operations")
        print ("\n Series values greater than the mean: %.4f"
        % Series1.mean())
        print (Series1 [Series1> Series1.mean()])
        print ("\n Series values greater than the
        Meadian:%.4f" % Series1.median())
        print (Series1 [Series1> Series1.median()])
        print ("\nExponential value ")
        Series1Exp = np.exp(Series1)
        print (Series1Exp)

        Series Sample operations

        Series values greater than the mean: 0.1766
        a    0.350241
        c    0.704124
        d    0.866934
        dtype: float64

        Series values greater than the Median: 0.5272
        c    0.704124
        d    0.866934
        dtype: float64
```

```
Exponential value
a     1.419409
b     0.296769
c     2.022075
d     2.379604
dtype: float64
```

Creating a Series from a Dictionary

You can create a series directly from a dictionary, as shown in Listing 3-30. If you don't explicitly pass the index, Python version +3.6 considers the series index by the dictionary insertion order. Otherwise, the series index will be the lexically ordered list of the dictionary keys.

Listing 3-30. Creating a Series from a Dictionary

```
In [12]: dict = {'m' : 2, 'y' : 2018, 'd' : 'Sunday'}
         print ("\nSeries of non declared index")
         SeriesDict1 = pd.Series(dict)
         print(SeriesDict1)
         print ("\nSeries of declared index")
         SeriesDict2 = pd.Series(dict, index=['y', 'm', 'd',
         's']) print(SeriesDict2)
         Series of non declared index
         d    Sunday
         m    2
         y    2018
         dtype: object

         Series of declared index
         y    2018
         m    2
```

```
d     Sunday
s     NaN
dtype: object
```

You can use the get method to access a series values by index label, as shown in Listing 3-31.

Listing 3-31. Altering a Series and Using the Get() Method

```
In [13]: print ("\nUse the get and set methods to access"
                "a series values by index label\n")
         SeriesDict2 = pd.Series(dict, index=['y', 'm', 'd',
         's']) print (SeriesDict2['y']) # Display the year
         SeriesDict2['y']=1999     # change the year value
         print (SeriesDict2)           # Display all dictionary
         values print (SeriesDict2.get('y')) # get specific
         value by its key
         Use the get and set methods to access a series values
         by index label
         2018
         y    1999
         m    2
         d    Sunday
         s    NaN
         dtype: object
         1999
```

Creating a Series from a Scalar Value

If data is a scalar value, an index must be provided. The value will be repeated to match the length of index. See Listing 3-32.

Listing 3-32. Creating a Series Using a Scalar Value

```
In [14]: print ("\n CREATE SERIES FORM SCALAR VALUE ")
         Scl = pd.Series(8., index=['a', 'b', 'c', 'd'])
         print (Scl)
         CREATE SERIES FORM SCALAR VALUE
         a   8.0
         b   8.0
         c   8.0
         d   8.0
         dtype: float64
```

Vectorized Operations and Label Alignment with Series

Series operations automatically align the data based on label. Thus, you can write computations without giving consideration to whether the series involved have the same labels. If labels are not matches, it gives a missing value NaN. See Listing 3-33.

Listing 3-33. Vectorizing Operations on a Series

```
In [16]: SerX = pd.Series([1,2,3,4], index=['a', 'b', 'c', 'd'])
          print ("Addition");
          print( SerX + SerX)
          print ("Addition with non-matched labels");
          print (SerX[1:] + SerX[:-1])
          print ("Multiplication");
          print (SerX * SerX)
          print ("Exponential");
          print (np.exp(SerX))
```

```
Addition
a 2
b 4
c 6
d 8
dtype: int64

Addition with non-matched labels
a   NaN
b   4.0
c   6.0
d   NaN
dtype: float64

Multiplication
a   1
b   4
c   9
d   16
dtype: int64

Exponential
a   2.718282
b   7.389056
c   20.085537
d   54.598150
dtype: float64
```

Name Attribute

You can name a series; also, you can alter a series, as shown in Listing 3-34.

Listing 3-34. Using a Series Name Attribute

```
In [17]:std = pd.Series([77,89,65,90], name='StudentsMarks')
        print (std.name)
        std = std.rename("Marks")
        print (std.name)
        StudentsMarks
        Marks
```

Data Frames

A *data frame* is a two-dimensional tabular labeled data structure with columns of potentially different types. A data frame can be created from numerous data collections such as the following:

- A 1D ndarray, list, dict, or series
- 2D Numpy ndarray
- Structured or record ndarray
- A series
- Another data frame

A data frame has arguments, which are an index (row labels) and columns (column labels).

Creating Data Frames from a Dict of Series or Dicts

You can simply create a data frame from a dictionary of series; it's also possible to assign an index. If there is an index without a value, it gives a NaN value, as shown in Listing 3-35.

Listing 3-35. Creating a Data Frame from a Dict of Series

```
In [5]: import pandas as pd
          dict1 = {'one' : pd.Series([1., 2., 3.],
          index=['a', 'b', 'c']),
          'two' : pd.Series([1., 2., 3., 4.],
          index=['a', 'b', 'c', 'd'])}
          df = pd.DataFrame(dict1)
          df
Out[5]: one  two
        a    1.0    1.0
        b    2.0    2.0
        c    3.0    3.0
        d    NaN    4.0

In [6]: # set index for the DataFrame
        pd.DataFrame(dict1, index=['d', 'b', 'a'])
Out[6]: one  two
        d    NaN    4.0
        b    2.0    2.0
        a    1.0    1.0

In [8]: # Control the labels appearance of the DataFrame
pd.DataFrame(dict1, index=['d', 'b', 'a'], columns=['two',
'three', 'one'])
Out[8]: two  three  one
        d    4.0    NaN    NaN
        b    2.0    NaN    2.0
        a    1.0    NaN    1.0
```

Creating Data Frames from a Dict of Ndarrays/Lists

When you create a data frame from an ndarray, the ndarrays must all be the same length. Also, the passed index should be of the same length as the arrays. If no index is passed, the result will be `range(n)`, where n is the array length. See Listing 3-36.

Listing 3-36. Creating a Data Frame from an Ndarray

```
In [11]: # without index
           ndarrdict = {'one' : [1., 2., 3., 4.],'two' :
           [4., 3., 2., 1.]}
          pd.DataFrame(ndarrdict)
Out[11]:           one    two
             0     1.0    4.0
             1     2.0    3.0
             2     3.0    2.0
             3     4.0    1.0

In [12]: # Assign index
         pd.DataFrame(ndarrdict, index=['a', 'b', 'c', 'd'])
Out[12]:     one    two
             a      1.0       4.0
             b      2.0       3.0
             c      3.0       2.0
             d      4.0       1.0
```

Creating Data Frames from a Structured or Record Array

Listing 3-37 creates a data frame by first specifying the data types of each column and then the values of each row. `('A', 'i4')` determines the column label and its data type as integers, `('B', 'f4')` determines the label as B and the data type as float, and finally `('C', 'a10')` assigns the label C and data type as a string with a maximum of ten characters.

Listing 3-37. Creating a Data Frame from a Record Array

```
In [18]:import pandas as pd
         import numpy as np
         data = np.zeros((2,), dtype=[('A', 'i4'),('B', 'f4'),
         ('C', 'a10')])
         data[:] = [(1,2.,'Hello'), (2,3.,"World")]
         pd.DataFrame(data)
Out[18]:     A      B      C
           0 1      2.0    b'Hello'
           1 2      3.0    b'World'

In [16]: pd.DataFrame(data, index=['First', 'Second'])
Out[16]:             A     B      C
         First       1     2.0    b'Hello'
         Second      2     3.0    b'World'

In [17]: pd.DataFrame(data, columns=['C', 'A', 'B'])
Out[17]:      C        A      B
         0    b'Hello' 1      2.0
         1    b'World' 2      3.0
```

Creating Data Frames from a List of Dicts

Also, you can create data frame from a list of dictionaries, as shown in Listing 3-38.

Listing 3-38. Creating a Data Frame from a List of Dictionaries

```
In [19]: data2 = [{'A ': 1, 'B ': 2}, {'A': 5, 'B': 10, 'C': 20}]
         pd.DataFrame(data2)
Out[19]: A  B       C
         0   1  2   NaN
         1   5  10  20.0
In [20]: pd.DataFrame(data2, index=['First', 'Second'])
Out[20]:           A   B     C
          First    1   2    NaN
          Second   5   10   20.0
In [21]: pd.DataFrame(data2, columns=['A', 'B'])
Out[21]:    A   B
         0  1   2
         1  5  10
```

Creating Data Frames from a Dict of Tuples

Another method to create a multi-indexed data frame is to pass a dictionary of tuples, as indicated in Listing 3-39.

Listing 3-39. Creating a Data Frame from a Dictionary of Tuples

```
In [22]: pd.DataFrame({('a', 'b'): {('A', 'B'): 1, ('A', 'C'): 2},
                       ('a', 'a'): {('A', 'C'): 3, ('A',
                       'B'): 4},
                       ('a', 'c'): {('A', 'B'): 5, ('A',
                       'C'): 6},
                       ('b', 'a'): {('A', 'C'): 7, ('A',
                       'B'): 8},
                       ('b', 'b'): {('A', 'D'): 9, ('A',
                       'B'): 10}})
```

```
Out[22]:
```

		a			b	
		a	b	c	a	b
A	B	4.0	1.0	5.0	8.0	10.0
	C	3.0	2.0	6.0	7.0	NaN
	D	NaN	NaN	NaN	NaN	9.0

Selecting, Adding, and Deleting Data Frame Columns

Once you have a data frame, you can simply add columns, remove columns, and select specific columns. Listing 3-40 demonstrates how to alter a data frame and its related operations.

Listing 3-40. Adding Columns and Making Operations on a Created Data Frame

```
In [25]: # DATAFRAME COLUMN SELECTION, ADDITION, DELETION
         ndarrdict = {'one' : [1., 2., 3., 4.], 'two' :
         [4., 3., 2., 1.]}
         df = pd.DataFrame(ndarrdict, index=['a', 'b', 'c', 'd'])
         df
```

```
Out[25]:
```

	one	two
a	1.0	4.0
b	2.0	3.0
c	3.0	2.0
d	4.0	1.0

```
In [26]: df['three'] = df['one'] * df['two'] # Add column
         df['flag'] = df['one'] > 2      # Add column
         df
```

Out[26]:

	one	two	three	flag
a	1.0	4.0	4.0	False
b	2.0	3.0	6.0	False
c	3.0	2.0	6.0	True
d	4.0	1.0	4.0	True

You can insert a scalar value to a data frame; it will naturally be propagated to fill the column. Also, if you insert a series that does not have the same index as the data frame, it will be conformed to the data frame's index. To delete a column, you can use the `del` or pop method, as shown in Listing 3-41.

Listing 3-41. Adding a Column Using a Scalar and Assigning to a Data Frame

```
In [27]: df['Filler'] = 'HCT'
         df['Slic'] = df['one'][:2]
         df
```

Out[27]:

	one	two	three	flag	Filler	Slic
a	1.0	4.0	4.0	False	HCT	1.0
b	2.0	3.0	6.0	False	HCT	2.0
c	3.0	2.0	6.0	True	HCT	NaN
d	4.0	1.0	4.0	True	HCT	NaN

```
In [28]:# Delet columns
        del df['two']
        Three = df.pop('three')
        df
```

Out[28]:

	one	flag	Filler	Slic
a	1.0	False	HCT	1.0
b	2.0	False	HCT	2.0
c	3.0	True	HCT	NaN
d	4.0	True	HCT	NaN

```
In [29]: df.insert(1, 'bar', df['one'])
         df
```

Out[29]:

	one	bar	flag	Filler	Slic
a	1.0	1.0	False	HCT	1.0
b	2.0	2.0	False	HCT	2.0
c	3.0	3.0	True	HCT	NaN
d	4.0	4.0	True	HCT	NaN

By default, columns get inserted at the end. However, you can use the `insert()` function to insert at a particular location in the columns, as shown previously.

Assigning New Columns in Method Chains

A data frame has an `assign()` method that allows you to easily create new columns that are potentially derived from existing columns. Also, you can change values of specific columns by altering the columns and making the necessary operations, as in column A in Listing 3-42.

Listing 3-42. Using the assign() Method to Add a Derived Column

```
In [54]: import numpy as np
           import pandas as pd
         df = pd.DataFrame({"A": [1, 2, 3], "B": [4, 5, 6]})
         df = df.assign(C=lambda x: x['A'] + x['B'])
         df = df.assign( D=lambda x: x['A'] + x['C'])
         df
```

Out[54]:

	A	B	C	D
0	1	4	5	6
1	2	5	7	9
2	3	6	9	12

```
In [55]: df = df.assign( A=lambda x: x['A'] *2)
         df
```

Out[55]:

	A	B	C	D
0	2	4	5	6
1	4	5	7	9
2	6	6	9	12

Indexing and Selecting Data Frames

Table 3-6 summarizes the data frame indexing and selection methods of columns and rows.

Table 3-6. *Data Frame Indexing and Selection Methods*

Operation	Syntax	Result
Select column	df[col]	Series
Select row by label	df.loc[label]	Series
Select row by integer location	df.iloc[loc]	Series
Slice rows	df[5:10]	Data frame
Select rows by Boolean vector	df[bool_vec]	Data frame

Listing 3-43 applies different approaches for rows and columns selections from a data frame.

Listing 3-43. *Data Frame Row and Column Selections*

```
In [56]: df

Out[56]:
      A  B  C  D
   0  2  4  5  6
   1  4  5  7  9
   2  6  6  9 12

In [61]: df['B']

Out[61]: 0    4
         1    5
         2    6
         Name: B, dtype: int64
```

```
In [59]: df.iloc[2]
```

```
Out[59]: A     6
         B     6
         C     9
         D    12
         Name: 2, dtype: int64
```

```
In [62]: df[1:]
```

Out[62]:

	A	B	C	D
1	4	5	7	9
2	6	6	9	12

```
In [65]: df[df['C']>7]
```

Out[65]:

	A	B	C	D
2	6	6	9	12

See Listing 3-44.

Listing 3-44. Operations on Data Frames

```
In [69]:df1 = pd.DataFrame({"A": [1, 2, 3], "B": [4, 5, 6]})
        df2 = pd.DataFrame({"A": [7, 4, 6], "B": [10, 4, 15]})
        print (df1)
        print()
        print(df2)
```

```
   A  B
0  1  4
1  2  5
2  3  6

   A   B
0  7  10
1  4   4
2  6  15
```

```
In [70]: df1 + df2
```

Out[70]:

	A	B
0	8	14
1	6	9
2	9	21

```
In [71]: df1-df2
```

Out[71]:

	A	B
0	-6	-6
1	-2	1
2	-3	-9

```
In [72]: df2 - df1.iloc[2]
```

Out[72]:

	A	B
0	4	4
1	1	-2
2	3	9

```
In [75]: df2
```

Out[75]:

	A	B
0	7	10
1	4	4
2	6	15

```
In [78]: df2*2+1
```

Out[78]:

	A	B
0	15	21
1	9	9
2	13	31

Transposing a Data Frame

You can transpose a data frame using the T operator, as shown in Listing 3-45.

Listing 3-45. Transposing a Data Frame

```
In [78]: df2
```

Out[80]:

	A	B
0	7	10
1	4	4
2	6	15

```
In [78]: df2[:].T
```

Out[79]:

	0	1	2
A	7	4	6
B	10	4	15

Data Frame Interoperability with Numpy Functions

You can implement matrix operations using the dot method on a data frame. For example, you can implement matrix multiplication as in Listing 3-46.

Listing 3-46. Matrix Multiplications

```
In [78]: df1
```

Out[81]:

	A	B
0	1	4
1	2	5
2	3	6

```
In [78]: df1.T.dot(df1)
```

Out[82]:

	A	B
A	14	32
B	32	77

Panels

A *panel* is a container for three-dimensional data; it's somewhat less frequently used by Python programmers.

A panel creation has three main attributes.

- `items`: axis 0; each item corresponds to a data frame contained inside
- `major_axis`: axis 1; it is the index (rows) of each of the data frames
- `minor_axis`: axis 2; it is the columns of each of the data frames

Creating a Panel from a 3D Ndarray

You can create a panel from a 3D ndarray with optional axis labels, as shown in Listing 3-47.

Listing 3-47. Creating a Panel from a 3D Ndarray

```
In [3]:import pandas as pd
        import numpy as np
        P1 = pd.Panel(np.random.randn(2, 5, 4), items=['Item1',
             'Item2'],major_axis=pd.date_range('10/05/2018',
             periods=5), minor_axis=['A', 'B', 'C', 'D'])
        P1
```

```
Out[3]: <class 'pandas.core.panel.Panel'>
        Dimensions: 2 (items) x 5 (major_axis) x 4 (minor_axis)
        Items axis: Item1 to Item2
        Major_axis axis: 2018-10-05 00:00:00 to 2018-10-09 00:00:00
        Minor_axis axis: A to D
```

Creating a Panel from a Dict of Data Frame Objects

You can create a panel from a dictionary of a data frame, as shown in Listing 3-48.

Listing 3-48. Creating a Panel from a Dictionary of Data Frames

```
In [4]: data = {'Item1' : pd.DataFrame(np.random.randn(4, 3)),
                'Item2' : pd.DataFrame(np.random.randn(4, 2))}
        P2 = pd.Panel(data)
        P2
```

```
Out[4]: <class 'pandas.core.panel.Panel'>
        Dimensions: 2 (items) x 4 (major_axis) x 3 (minor_axis)
        Items axis: Item1 to Item2
        Major_axis axis: 0 to 3
        Minor_axis axis: 0 to 2
```

```
In [5]: p3 = pd.Panel.from_dict(data, orient='minor')
          p3
```

```
Out[5]: <class 'pandas.core.panel.Panel'>
        Dimensions: 3 (items) x 4 (major_axis) x 2 (minor_axis)
        Items axis: 0 to 2
        Major_axis axis: 0 to 3
        Minor_axis axis: Item1 to Item2
```

See Listing 3-49.

Listing 3-49. Creating a Panel from a Data Frame

```
In [26]: df = pd.DataFrame({'Item': ['TV', 'Mobile', 'Laptop'],
         'Price': np.random.randn(3)**2*1000})
           df
```

Out[26]:

	Item	Price
0	TV	3704.932147
1	Mobile	1348.142561
2	Laptop	336.985518

```
In [29]: data = {'stock1': df, 'stock2': df}
         panel = pd.Panel.from_dict(data, orient='minor')
         panel['Item']
```

Out[29]:

	stock1	stock2
0	TV	TV
1	Mobile	Mobile
2	Laptop	Laptop

```
In [30]: panel['Price']
```

Out[30]:

	stock1	stock2
0	3704.932147	3704.932147
1	1348.142561	1348.142561
2	336.985518	336.985518

Selecting, Adding, and Deleting Items

A panel is like a dict of data frames; you can slice elements, select items, and so on. Table 3-7 gives three operations for panel items selections.

Table 3-7. *Panel Item Selection and Slicing Operations*

Operation	Syntax	Result
Select item	`wp[item]`	Data frame
Get slice at `major_axis` label	`wp.major_xs(val)`	Data frame
Get slice at `minor_axis` label	`wp.minor_xs(val)`	Data frame

See Listing 3-50.

Listing 3-50. Slicing and Selecting Items from a Panel

```
In [33]: import pandas as pd
           import numpy as np
           P1 = pd.Panel(np.random.randn(2, 5, 4),
           items=['Item1',
                 'Item2'], major_axis=pd.date_
                 range('10/05/2018',
                  periods=5), minor_axis=['A', 'B', 'C', 'D'])
P1['Item1']
```

Out[33]:

	A	B	C	D
2018-10-05	-0.794656	1.082396	-0.368632	0.360976
2018-10-06	-0.281474	0.070584	-0.012636	-0.388089
2018-10-07	1.653752	0.487939	1.838114	-0.832078
2018-10-08	-0.145535	1.856141	0.107239	0.462018
2018-10-09	-0.816565	2.195793	-0.871674	-1.226616

```
In [34]: P1.major_xs(P1.major_axis[2])
```

Out[34]:

	Item1	Item2
A	1.653752	-0.496110
B	0.487939	0.990550
C	1.838114	1.492156
D	-0.832078	-0.197148

```
In [35]: P1.minor_axis
Out[35]: Index(['A', 'B', 'C', 'D'], dtype='object')

In [36]: P1.minor_xs('C')
```

Out[36]:

	Item1	Item2
2018-10-05	-0.368632	-0.989085
2018-10-06	-0.012636	0.266520
2018-10-07	1.838114	1.492156
2018-10-08	0.107239	-0.555847
2018-10-09	-0.871674	-0.468046

Summary

This chapter covered data collection structures in Python and their implementations. Here's a recap of what was covered:

- How to maintain a collection of data in different forms
- How to create lists and how to manipulate list content
- What a dictionary is and the purpose of creating a dictionary as a data container

- How to create tuples and what the difference is between tuple data structure and dictionary structure, as well as the basic tuple operations
- How to create a series from other data collection forms
- How to create data frames from different data collection structures and from another data frame
- How to create a panel as a 3D data collection from a series or data frame

The next chapter will cover file I/O processing and using regular expressions as a tool for data extraction and much more.

Exercises and Answers

1. Write a program to create a list of names; then define a function to display all the elements in the received list. Call the function to execute its statements and display all names in the list.

Answer:

```
In [124]: Students =["Ahmed", "Ali", "Salim", "Abdullah",
"Salwa"]
          def displaynames (x):
              for name in x:
                  print (name)
         displaynames(Students)   # Call the function display
         names
         Ahmed
         Ali
         Salim
```

```
Abdullah
Salwa
```

2. Write a program to read text file data and create a dictionary of all keywords in the text file. The program should count how many times each word is repeated inside the text file and then find the keyword with a highest repeated number. The program should display both the keywords dictionary and the most repeated word.

Answer:

```
In [4]: # read data from file and add it to dictionary for
processing
```

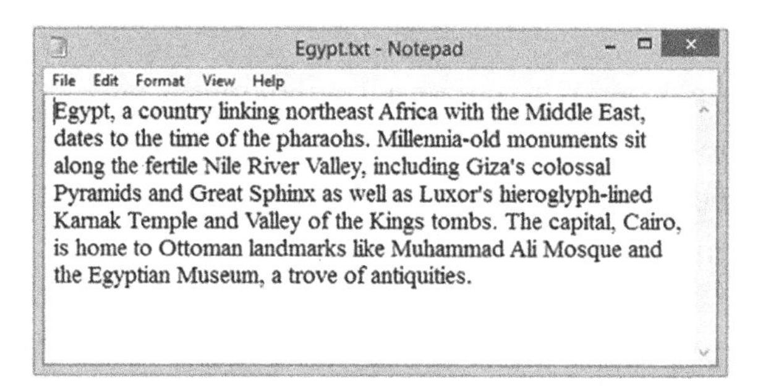
Egypt.txt - Notepad
File Edit Format View Help
Egypt, a country linking northeast Africa with the Middle East, dates to the time of the pharaohs. Millennia-old monuments sit along the fertile Nile River Valley, including Giza's colossal Pyramids and Great Sphinx as well as Luxor's hieroglyph-lined Karnak Temple and Valley of the Kings tombs. The capital, Cairo, is home to Ottoman landmarks like Muhammad Ali Mosque and the Egyptian Museum, a trove of antiquities.

```
handle = open("Egypt.txt")
text = handle.read()
words = text.split()

counts = dict()
for word in words:
        counts[word] = counts.get(word,0) + 1
print (counts)
bigcount = None
bigword = None
```

```
for word,count in counts.items():
    if bigcount is None or count > bigcount:
       bigword = word
       bigcount = count
print ("\n bigword and bigcount")
print (bigword, bigcount)
```

```
{'Egypt,': 1, 'a': 2, 'country': 1, 'linking': 1, 'northeast': 1, 'Africa': 1, 'with': 1, 'the': 6, 'Middle': 1, 'Eas
t,': 1, 'dates': 1, 'to': 2, 'time': 1, 'of': 3, 'pharaohs.': 1, 'Millennia-old': 1, 'monuments': 1, 'sit': 1, 'along
': 1, 'fertile': 1, 'Nile': 1, 'River': 1, 'Valley,': 1, 'including': 1, "Giza's": 1, 'colossal': 1, 'Pyramids': 1, '
and': 3, 'Great': 1, 'Sphinx': 1, 'as': 2, 'well': 1, "Luxor's": 1, 'hieroglyph-lined': 1, 'Karnak': 1, 'Temple': 1,
'Valley': 1, 'Kings': 1, 'tombs.': 1, 'The': 1, 'capital,': 1, 'Cairo,': 1, 'is': 1, 'home': 1, 'Ottoman': 1, 'landma
rks': 1, 'like': 1, 'Muhammad': 1, 'Ali': 1, 'Mosque': 1, 'Egyptian': 1, 'Museum,': 1, 'trove': 1, 'antiquities.': 1}

 bigword and  bigcount
the 6
```

3. Write a program to compare tuples of integers and tuples of strings.

Answer:

```
In [14]: print ((100, 1, 2) > (150, 1, 2))
         print ((0, 1, 120) < (0, 3, 4))
         print (( 'Javed', 'Salwa' ) > ('Omar', 'Sam'))
         print (( 'Khalid', 'Ahmed') < ('Ziad', 'Majid'))
         False
         True
         False
         True
```

4. Write a program to create a series to maintain three students' names and GPA values.

Name	GPA
Omar	2.5
Ali	3.5
Osama	3

Answer:

```
In [41]: data = {'Omar' : 2.5, 'Ali' : 3.5, 'Osama' : 3.0}
         pd.Series(data)

Out[41]: Ali      3.5
         Omar     2.5
         Osama    3.0
         dtype: float64

In [42]: pd.Series(data, index=['Ali', 'Omar', 'Osama'])

Out[42]: Ali      3.5
         Omar     2.5
         Osama    3.0
         dtype: float64
```

5. Write a program to create a data frame to maintain three students' names associated with their grades in three courses and then add a new column named Mean to maintain the calculated mean mark per course. Display the final data frame.

Name	Course 1	Course2	Course3
Omar	90	50	89
Ali	78	75	73
Osama	67	85	80

Answer:

```
In [31]: data = {'Omar': [90, 50, 89], 'Ali': [78, 75, 73],
'Osama': [67, 85, 80]}
          df1 = pd.DataFrame (data, index= ['Course1',
          'Course2', 'Course3'])
          df1
```

Out[51]:

	Ali	Omar	Osama
Course1	78	90	67
Course2	75	50	85
Course3	73	89	80

```
In [32]: df1['Omar']

Out[32]:Course1     90
         Course2     50
         Course3     89
         Name: Omar, dtype: int64

In [33]: df1['Mean'] = (df1['Ali'] + df1['Omar'] +
df1['Osama'])/3
      df1
```

Out[54]:

	Ali	Omar	Osama	Mean
Course1	78	90	67	78.333333
Course2	75	50	85	70.000000
Course3	73	89	80	80.666667

CHAPTER 4

File I/O Processing and Regular Expressions

In this chapter, you'll study input-output functions and file processing. In addition, you'll study regular expressions and how to extract data that matches specific patterns.

File I/O Processing

Python provides numerous methods for input, output, and file processing. You can get input from the screen and output data to the screen as well as read data from files and store data in files.

Data Input and Output

You can read data from a user using the `input()` function. Received data by default is in text format. Hence, you should use conversion functions to convert the data into numeric values if required, as shown in Listing 4-1.

O. Embarak, *Data Analysis and Visualization Using Python*,
https://doi.org/10.1007/978-1-4842-4109-7_4

Listing 4-1. Screen Data Input/Output

```
In [2]: Name = input ("Enter your name: ")
        Name
Enter your name: Osama Hashim
Out[2]: 'Osama Hashim'

In [3]: Mark = input("Enter your mark: ") Mark = float(Mark)
Enter your mark: 92
In [4]:print ("Welcome to Grading System \nHCT 2018")
        print ("\nCampus\t Name\t\tMark\tGrade")
        if (Mark>=85):
            Grade="B+"
            print ("FMC\t", Name,"\t",Mark,"\t", Grade)
Welcome to Grading System
HCT 2018
Campus       Name         Mark    Grade
FMC       Osama Hashim    92.0    B+
```

Here you are converting the `Mark` value into a float using `float(Mark)`. You use `\t` to add tabs and `\n` to jump lines on the screen.

Opening and Closing Files

Python's built-in `open()` function is used to open a file stored on a computer hard disk or in the cloud. Here's its syntax:

```
file object = open(file_name [, access_mode][, buffering])
```

Table 4-1 describes its modes.

Table 4-1. *Open File Modes*

No.	Modes	Description
1	r	Opens a file for reading only; the default mode
2	rb	Opens a file for reading only in binary format
3	r+	Opens a file for both reading and writing
4	rb+	Opens a file for both reading and writing in binary format
5	w	Opens a file for writing only
6	wb	Opens a file for writing only in binary format
7	w+	Opens a file for both writing and reading
8	wb+	Opens a file for both writing and reading in binary format
9	a	Opens a file for appending
10	ab	Opens a file for appending in binary format
11	a+	Opens a file for both appending and reading
12	ab+	Opens a file for both appending and reading in binary format

File Object Attributes

Python provides various methods for detecting the open file's information, as shown in Table 4-2.

Table 4-2. *Opened File Attributes*

No.	Attribute	Description
1	`file.closed`	Returns true if the file is closed; false otherwise
2	`file.mode`	Returns access mode with which file was opened
3	`file.name`	Returns name of the file

Listing 4-2 displays the attributes of an open file called `Egypt.txt`.

Listing 4-2. Opened File Attributes

```
In [41]: # Open a file and find its attributes
         Filehndl = open("Egypt.txt", "r")
         print ("Name of the file: ", Filehndl.name)
         print ("Closed or not : ", Filehndl.closed)
         print ("Opening mode : ", Filehndl.mode)
Name of the file: Egypt.txt
Closed or not : False
Opening mode : r
```

You can close an opened file using the `close()` method to clear all related content from memory and to close any opened streams to the back-end file, as shown in Listing 4-3.

Listing 4-3. Closing Files

```
In [40]: Filehndl = open("Egypt.txt", "r")
         print ("Closed or not : ", Filehndl.closed)
         Filehndl.close()
         print ("Closed or not : ", Filehndl.closed)
Closed or not : False
Closed or not : True
```

Reading and Writing to Files

The `file.write()` method is used to write to a file as shown in below figure, and the `file.read()` method is used to read data from an opened file. A file can be opened for writing (`W`), reading (`r`), or both (`r+`), as shown in Listing 4-4.

Listing 4-4. Writing to a File

```
In [39]:Filehndl = open("Egypt.txt", "w+")
        Filehndl.write( "Python Processing Files\nMay
        2018!!\n")
        # Close opend file
        Filehndl.close()
```

As shown in the following figure, data has been written into the "Egypt.txt" file.

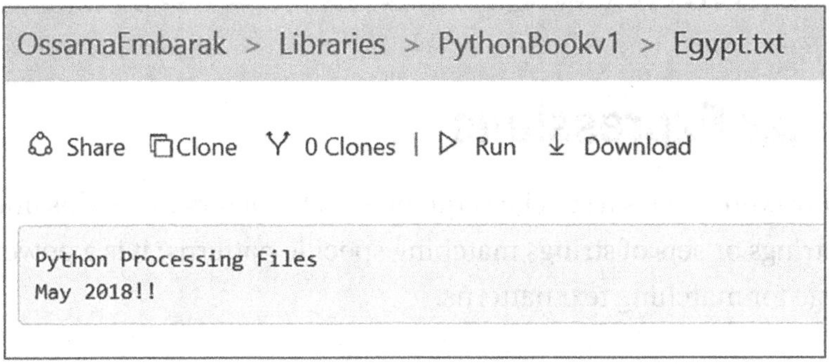

OssamaEmbarak > Libraries > PythonBookv1 > Egypt.txt

Share Clone 0 Clones | Run Download

```
Python Processing Files
May 2018!!
```

The `rename()` method is used to rename a file; it takes two arguments: the current filename and the new filename. Also, the `remove()` method can be used to delete files by supplying the name of the file to be deleted as an argument.

```
In [34]: import os
         os.rename( "Egypt.txt", "test2.txt" )
         os.remove( "test2.txt" )
```

Directories in Python

Python provides various methods for creating and accessing directories. Listing 4-5 demonstrates how to create, move, and delete directories. You can find the current working directory using Python's `getcwd()` method.

Listing 4-5. Creating and Deleting Directories

```
In [35]: import os
         os.mkdir("Data 1") # create a directory
         os.mkdir("Data_2")
         os.chdir("Data_3")     # create a Childe directory
         os.getcwd()            # Get the current working
                                  directory

         os.rmdir('Data 1') # remove a directory
         os.rmdir('Data_3') # remove a directory
```

Regular Expressions

A *regular expression* is a special sequence of characters that helps find other strings or sets of strings matching specific patterns; it is a powerful language for matching text patterns.

Regular Expression Patterns

Different regular expression syntax can be used for extracting data from text files, XML, JSON, HTML containers, and so on.

Table 4-3 lists some Python regular expression syntax.

***Table 4-3.** Python Regular Expression Syntax*

No.	Pattern	Description
1	`^`	Matches beginning of the line.
2	`$`	Matches end of the line.
3	`.`	Matches any single character except a newline.
4	`[...]`	Matches any single character in brackets.
5	`[^...]`	Matches any single character not in brackets.
6	`re*`	Matches zero or more occurrences of the preceding expression.
7	`re+`	Matches one or more occurrence of the preceding expression.
8	`re?`	Matches zero or one occurrence of the preceding expression.
9	`re{ n}`	Matches exactly *n* number of occurrences of the preceding expression.
10	`re{ n,}`	Matches *n* or more occurrences of the preceding expression.
11	`re{ n, m}`	Matches at least *n* and at most *m* occurrences of the preceding expression.
12	`a| b`	Matches either *a* or *b*.
13	`(re)`	Groups regular expressions and remembers matched text.
14	`(?imx)`	Temporarily toggles on *i*, *m*, or *x* options within a regular expression.
15	`(?-imx)`	Temporarily toggles off *i*, *m*, or *x* options within a regular expression.
16	`(?: re)`	Groups regular expressions without remembering matched text.
17	`(?imx: re)`	Temporarily toggles on *i*, *m*, or *x* options within parentheses.

(continued)

Table 4-3. *(continued)*

No.	Pattern	Description
18	(?-imx: re)	Temporarily toggles off *i*, *m*, or *x* options within parentheses.
19	(?#...)	Comment.
20	(?= re)	Specifies the position using a pattern. Doesn't have a range.
21	(?! re)	Specifies the position using pattern negation. Doesn't have a range.
22	(?> re)	Matches independent pattern without backtracking.
23	\w	Matches word characters.
24	\W	Matches nonword characters.
25	\s	Matches whitespace. Equivalent to [\t\n\r\f].
26	\S	Matches nonwhitespace.
27	\d	Matches digits. Equivalent to [0-9].
28	\D	Matches nondigits.
29	\A	Matches beginning of the string.
30	\Z	Matches end of the string. If a newline exists, it matches just before the newline.
31	\z	Matches end of the string.
32	\G	Matches point where the last match finished.
33	\b	Matches word boundaries when outside brackets.
34	\B	Matches nonword boundaries.
35	\n, \t, etc.	Matches newlines, carriage returns, tabs, etc.
36	\1...\9	Matches nth grouped subexpression.
37	\10	Matches nth grouped subexpression if it matched already.

For instance, if you have a text file of e-mail log data and you want to extract only the text lines where the `@uct.ac.za` pattern appears, then you can use iteration to capture only the lines with the given pattern, as shown in Listing 4-6.

Listing 4-6. Reading and Processing a Text File

```
In [46]: print ("\nUsing in to select lines // only print lines
which has specific string ")
          fhand = open('Emails.txt')
          for line in fhand:
              line = line.rstrip()

              if not '@uct.ac.za' in line :
                      continue
              print (line)
```

You can extract only the lines starting with `From:`. Once it has been extracted, then you can split each line into a list and slice only the e-mail element, as indicated in Listing 4-7 and Listing 4-8.

Listing 4-7. Extracting Lines Starting with a Specific Pattern

```
In [45]: print("\nSearching Through a File\n")
         fhand = open('Emails.txt')
         for line in fhand:
              line = line.rstrip()
              if line.startswith('From:') :
                 print (line)
          Searching Through a File
          From: stephen.marquard@uct.ac.za
          From: louis@media.berkeley.edu
          From: zqian@umich.edu
          From: rjlowe@iupui.edu
```

```
From: zqian@umich.edu
From: rjlowe@iupui.edu
From: cwen@iupui.edu
From: cwen@iupui.edu
From: gsilver@umich.edu
From: gsilver@umich.edu
From: zqian@umich.edu
From: gsilver@umich.edu
From: wagnermr@iupui.edu
From: zqian@umich.edu
From: antranig@caret.cam.ac.uk
From: gopal.ramasammycook@gmail.com
From: david.horwitz@uct.ac.za
From: david.horwitz@uct.ac.za
From: david.horwitz@uct.ac.za
From: david.horwitz@uct.ac.za
From: stephen.marquard@uct.ac.za
From: louis@media.berkeley.edu
From: louis@media.berkeley.edu
From: ray@media.berkeley.edu
From: cwen@iupui.edu
From: cwen@iupui.edu
From: cwen@iupui.edu
```

Listing 4-8. Extracting e-mails without regular expressions

```
In [47]: print("\nSearching Through a File\n") fhand =
         open('Emails.txt')
          for line in fhand:
             line = line.rstrip()
             if line.startswith('From:') :
                 line = line.split()
                  print (line[1])
```

```
Searching Through a File
stephen.marquard@uct.ac.za louis@media.berkeley.edu
zqian@umich.edu
rjlowe@iupui.edu
zqian@umich.edu
rjlowe@iupui.edu
cwen@iupui.edu
cwen@iupui.edu
gsilver@umich.edu
gsilver@umich.edu
zqian@umich.edu
gsilver@umich.edu
wagnermr@iupui.edu
zqian@umich.edu
antranig@caret.cam.ac.uk
gopal.ramasammycook@gmail.com
david.horwitz@uct.ac.za
david.horwitz@uct.ac.za
david.horwitz@uct.ac.za
david.horwitz@uct.ac.za
stephen.marquard@uct.ac.za
louis@media.berkeley.edu
louis@media.berkeley.edu
ray@media.berkeley.edu
cwen@iupui.edu
cwen@iupui.edu
cwen@iupui.edu
```

Although regular expressions are useful for extracting data from word bags, they should be carefully used. The regular expression in Listing 4-9 finds all the text starting with a capital *X* followed by any character repeated zero or more times and ending with a colon (:).

Listing 4-9. Regular Expression Example

```
In [48]: import re
         print ("\nRegular Expressions\n'^X.*:' \n") hand =
         open('Data.txt')
         for line in hand:
              line = line.rstrip()
              y = re.findall('^X.*:',line)
              print (y)
```

This is a text file maintaining text data which we used to apply regular expressions as shown below.

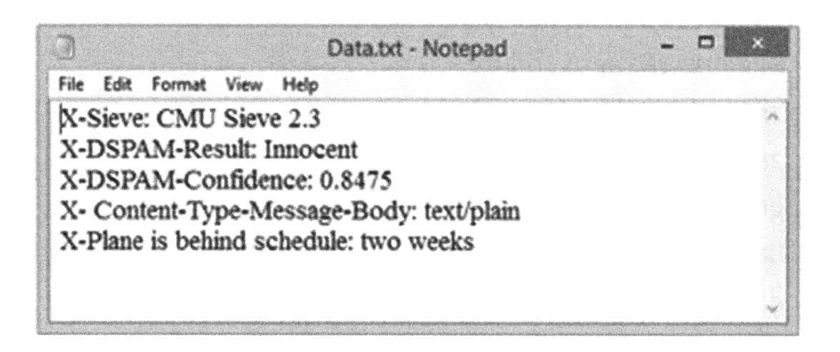

In the following code, the expression `'^X.*:'` retrieves all lines starting with a capital *X* followed by any character including white spaces zero or more times and ending with a colon delimiter (:) . However, it doesn't consider the whitespaces. Listing 4-10 retrieves only the values that have no whitespaces included in the matched patterns.

```
'X.*:'
['X-Sieve:']
['X-DSPAM-Result:']
['X-DSPAM-Confidence:']
['X- Content-Type-Message-Body:']
['X-Plane is behind schedule:']
```

Listing 4-10. Extracting Nonwhitespace Patterns

```
In [49]: print ("\nRegular Expressions\nWild-Card Characters
'^X-\S+:'\n")
         hand = open('Data.txt')
         for line in hand:
              line = line.rstrip()
              y = re.findall('^X-\S+:',line) # match any
              nonwhite space characters
              print (y)
              Regular Expressions
              Wild-Card Characters 'X-\S+:'
              ['X-Sieve:']
              ['X-DSPAM-Result:']
              ['X-DSPAM-Confidence:']
              []
              []
```

Regular expressions enable you to extract numerical values within a string and find specific patterns of characters within a string of characters, as shown in Listing 4-11.

Listing 4-11. Extracting Numerical Values and Specific Characters

```
In [50]: print ("\n Matching and Extracting Data \n")
         x = 'My 2 favorite numbers are 19 and 42'
         y = re.findall('[0-9]+',x)
         print (y)
         Matching and Extracting Data
         ['2', '19', '42']
```

```
In [51]: y = re.findall('[AEsOUn]+',x) # find any of these
characters in string
         print (y)
         ['n', 's', 'n']
```

Although regular expressions are useful for extracting data, they should be carefully implemented. The following examples show the greedy and nongreedy extraction. In the first example in Listing 4-12, Python finds a string starting with *F* and containing any number of characters up to a colon and then stops when it reaches the end of the line. That is why it continues to retrieve characters even when it finds the first colon. In the second example, `re.findall('^F.+?:', x)` asks Python to retrieve characters starting with an *F* and ending with the first occurrence of a delimiter, which is a colon regardless of whether it reached the end of the line or not.

Listing 4-12. Greedy and Nongreedy Matching

```
In [52]: print ("\nGreedy Matching \n")
         x = 'From: Using the : character'
         y = re.findall('^F.+:', x)
         print (y)
         Greedy Matching
         ['From: Using the :']
In [53]: print ("\nNon-Greedy Matching \n")
         x = 'From: Using the : character'
         y = re.findall('^F.+?:', x)
         print (y)
         Non-Greedy Matching
         ['From:']
```

Table 4-4 demonstrates various implementations of regular expressions.

Table 4-4. *Examples of Regular Expressions*

No.	Example	Description
1	`[Pp]ython`	Matches "`Python`" or "`python`"
2	`rub[ye]`	Matches "`ruby`" or "`rube`"
3	`[aeiou]`	Matches any one lowercase vowel
4	`[0-9]`	Matches any digit; same as `[0123456789]`
5	`[a-z]`	Matches any lowercase ASCII letter
6	`[A-Z]`	Matches any uppercase ASCII letter
7	`[a-zA-Z0-9]`	Matches any of the above
8	`[^aeiou]`	Matches anything other than a lowercase vowel
9	`[^0-9]`	Matches anything other than a digit

Special Character Classes

Some special characters are used within regular expressions to extract data. Table 4-5 summarizes some of these special characters.

Table 4-5. *Regular Expression Special Characters*

No.	Example	Description
1	.	Matches any character except newline
2	\d	Matches a digit: `[0-9]`
3	\D	Matches a nondigit: `[^0-9]`
4	\s	Matches a whitespace character: `[ \t\r\n\f]`
5	\S	Matches nonwhitespace: `[^ \t\r\n\f]`
6	\w	Matches a single word character: `[A-Za-z0-9_]`
7	\W	Matches a nonword character: `[^A-Za-z0-9_]`

Repetition Classes

It is possible to have a string with different spelling such as "ok" and "okay." To handle such cases, you can use repetition expressions, as shown in Table 4-6.

Table 4-6. *Regular Expression Repetition Characters*

No.	Example	Description
1	`ruby?`	Matches "rub" or "ruby"; the *y* is optional
2	`ruby*`	Matches "rub" plus zeros or more *y*s
3	`ruby+`	Matches "rub" plus one or more *y*s
4	`\d{3}`	Matches exactly three digits
5	`\d{3,}`	Matches three or more digits
6	`\d{3,5}`	Matches three, four, or five digits

Alternatives

Alternatives refer to expressions where you can use multiple expression statements to extract data, as shown in Table 4-7.

Table 4-7. *Alternative Regular Expression Characters*

No	Example	Description
1	`python|RLang`	Matches "python" or " RLang "
2	`R(L|Lang))`	Matches " RL" or " RLang"
3	`Python(!+|\?)`	"Python" followed by one or more ! or one ?

Anchors

Anchors enable you to determine the position in which you can find the match pattern in a string. Table 4-8 demonstrates numerous examples of anchors.

Table 4-8. *Anchor Characters*

No.	Example	Description
1	`^Python`	Matches "Python" at the start of a string or internal line
2	`Python$`	Matches "Python" at the end of a string or line
3	`\APython`	Matches "Python" at the start of a string
4	`Python\Z`	Matches "Python" at the end of a string
5	`\bPython\b`	Matches "Python" at a word boundary
6	`\brub\B`	\B is nonword boundary: matches "rub" in *rube* and *ruby* but not on its own
7	`Python(?=!)`	Matches "Python," if followed by an exclamation point
8	`Python(?!!)`	Matches "Python," if not followed by an exclamation point

Not only are regular expressions used to extract data from strings, but various built-in methods can be used for the same purposes. Listing 4-13 demonstrates the use of methods versus regular expressions to extract the same characters.

Listing 4-13. The Use of Methods vs. Regular Expressions

```
In [54]: import re
         print ("\nFine-Tuning String Extraction \n")
         mystr="From ossama.embarak@hct.ac.ae Sat Jun 5
         08:14:16 2018" Extract = re.findall('\S+@\S+',mystr)
```

```
        print (Extract)
        E_xtracted = re.findall('^From.*? (\S+@\S+)',mystr) #
        non greedy white space
        print (E_xtracted)
        print (E_xtracted[0])
        Fine-Tuning String Extraction

        ['ossama.embarak@hct.ac.ae']
        ['ossama.embarak@hct.ac.ae']
ossama.embarak@hct.ac.ae

In [57]: mystr="From ossama.embarak@hct.ac.ae Sat Jun 5
08:14:16 2018"
        atpos = mystr.find('@')
        sppos = mystr.find(' ',atpos) # find white space
        starting from atpos
        host = mystr[atpos+1 : sppos]
        print (host)
        usernamepos = mystr.find(' ')
        username = mystr[usernamepos+1 : atpos]
        print (username)
        hct.ac.ae
        ossama.embarak
```

`re.findall('@([^ ]*)',mystr)` retrieves a substring in the `mystr` string, which starts after @and continues until finding the whitespace. Similarly, `re.findall('^From .*@([^ ]*)', mystr)` retrieves a substring in the `mystr` string, which starts after `From` and finds zero or more characters and then the @ character *and then anything other than whitespace characters.* See Listing 4-14.

Listing 4-14. Using the Regular Expression findall() Method

```
In [58]: print ("\n The Regex Version\n")
         import re
         mystr="From ossama.embarak@hct.ac.ae Sat Jun 5
         08:14:16 2018"
         Extract = re.findall('@([^ ]*)',mystr)
         print (Extract)
         Extract = re.findall('^From .*@([^ ]*)',mystr)
         print (Extract)
         The Regex Version
         ['hct.ac.ae']
         ['hct.ac.ae']
In [59]: print ("\nScape character \n")
         mystr = 'We just received $10.00 for cookies and
         $20.23 for juice'
         Extract = re.findall('\$[0-9.]+',mystr)
         print (Extract)
         Scape character
         ['$10.00', '$20.23']
```

Summary

This chapter covered input/output data read or pulled from stored files or directly read from users. Let's recap what was covered in this chapter.

- The chapter covered how to open files for reading, writing, or both. Furthermore, it covered how to access the attributes of open files and close all opened sessions.
- The chapter covered how to collect data directly for users via the screen.

- It covered regular expressions and their patterns and special character usage.
- The chapter covered how to apply regular expressions to extract data and how to use alternatives, anchors, and repetition expressions for data extraction.

The next chapter will study techniques of gathering and cleaning data for further processing, and much more.

Exercises and Answer

1. Write a Python script to extract a course number, code, and name from the following text using regular expressions:

```
CoursesData = """101 COM Computers
205 MAT Mathematics
189 ENG English"""
```

Answer:

```
In [60]: import re
         CoursesData = """101 COM Computers
                                     205 MAT  Mathematics
                                     189 ENG  English"""
In [61]: # Extract all course numbers
         Course_numbers = re.findall('[0-9]+', CoursesData)

         print (Course_numbers)

         # Extract all course codes
         Course_codes = re.findall('[A-Z]{3}', CoursesData)

         print (Course_codes)
```

```
# Extract all course names
Course_names = re.findall('[A-Za-z]{4,}', CoursesData)
print (Course_names)

['101', '205', '189']
['COM', 'MAT', 'ENG']
['Computers', 'Mathematics', 'English']
```

2. Write a Python script to extract each course's details in a tuple form from the following text using regular expressions. In addition, use regular expressions to retrieve string values in the `CoursesData` and then retrieve numerical values in `CoursesData`.

Answer:

```
CoursesData = """101 COM Computers
205 MAT Mathematics
189 ENG English"""
```

In [63]: # define the course text pattern groups and extract

```
course_pattern = '([0-9]+)\s*([A-Z]{3})\s*([A-Za-z]
{4,})'
re.findall(course_pattern, CoursesData)
Out[63]: [('101', 'COM', 'Computers'),
 ('205', 'MAT', 'Mathematics'),
 ('189', 'ENG', 'English')]

In [64]: print(re.findall('[a-zA-Z]+', CoursesData)) # []
Matches any character inside
['COM', 'Computers', 'MAT', 'Mathematics', 'ENG', 'English']

In [65]: print(re.findall('[0-9]+', CoursesData)) # [] Matches
any numeric inside
['101', '205', '189']
```

3. Write a Python script to extract digits of size 4 and digits of size 2 to 4 using regular expressions.

Answer:

```
CoursesData = """101 COM Computers
205 MAT Mathematics
189 ENG English"""
```

```
In [66]: import re
         CoursesData = """10 COM Computers

                          205 MAT Mathematics 1899 ENG English"""
         print(re.findall('\d{4}', CoursesData)) # {n} Matches
         repeat n times.
         print(re.findall('\d{2,4}', CoursesData))

         ['1899']
         ['10', '205', '1899']
```

CHAPTER 5

Data Gathering and Cleaning

In the 21st century, data is vital for decision-making and developing long-term strategic plans. Python provides numerous libraries and built-in features that make it easy to support data analysis and processing. Making business decisions, forecasting weather, studying protein structures in biology, and designing a marketing campaign are all examples that require collecting data and then cleaning, processing, and visualizing it.

There are five main steps for data science processing.

1. *Data acquisition* is where you read data from various sources of unstructured data, semistructured data, or full-structured data that might be stored in a spreadsheet, comma-separated file, web page, database, etc.

2. *Data cleaning* is where you remove noisy data and make operations needed to keep only the relevant data.

3. *Exploratory analysis* is where you look at your cleaned data and make statistical processing fits for specific analysis purposes.

O. Embarak, *Data Analysis and Visualization Using Python*,
https://doi.org/10.1007/978-1-4842-4109-7_5

4. An *analysis model* needs to be created. Advanced tools such as machine learning algorithms can be used in this step.

5. *Data visualization* is where the results are plotted using various systems provided by Python to help in the decision-making process.

Python provides several libraries for data gathering, cleaning, integration, processing, and visualizing.

- Pandas is an open source Python library used to load, organize, manipulate, model, and analyze data by offering powerful data structures.
- Numpy is a Python package that stands for "numerical Python. It is a library consisting of multidimensional array objects and a collection of routines for manipulating arrays. It can be used to perform mathematical, logical, and linear algebra operations on arrays.
- SciPy is another built-in Python library for numerical integration and optimization.
- Matplotlib is a Python library used to create 2D graphs and plots. It supports a wide variety of graphs and plots such as histograms, bar charts, power spectra, error charts, and so on, with additional formatting such as control line styles, font properties, formatting axes, and more.

Cleaning Data

Data is collected and entered manually or automatically using various methods such as weather sensors, financial stock market data servers, users' online commercial preferences, etc. Collected data is not error-free and usually has various missing data points and erroneously entered data. For instance, online users might not want to enter their information because of privacy concerns. Therefore, treating missing and noisy data (NA or NaN) is important for any data analysis processing.

Checking for Missing Values

You can use built-in Python methods to check for missing values. Let's create a data frame using the Numpy and Pandas libraries. Include the index values *a* to *h*, and give the columns labels of `stock1`, `stock2`, and `stock3`, as shown in Listing 5-1.

Listing 5-1. Creating a Data Frame Including NaN

```
In [47]: import pandas as pd
         import numpy as np
         dataset = pd.DataFrame(np.random.randn(5, 3),
         index=['a', 'c', 'e', 'f', 'h'],columns=['stock1',
         'stock2', 'stock3'])
         dataset.rename(columns={"one":'stock1',"two":'stock2',
         "three":'stock3'}, inplace=True)
         dataset = dataset.reindex(['a', 'b', 'c', 'd', 'e',
         'f', 'g', 'h'])
         print (dataset)
```

```
     stock1    stock2    stock3
a -0.716435  0.646375  0.403254
b       NaN       NaN       NaN
c  0.923383 -0.354701 -0.594661
d       NaN       NaN       NaN
e  1.039185  0.984489  0.902545
f -0.398857 -0.205501 -1.859085
g       NaN       NaN       NaN
h  0.228843  0.049838  0.400659
```

It should be clear that you can use Numpy to create an array of random values, as shown in Listing 5-2.

Listing 5-2. Creating a Matrix of Random Values

```
In [46]: import numpy as np
         np.random.randn(5, 3)

Out[4]: array([[-2.45374913,  1.26130579, -1.09523564],
              [-0.00900845, -1.23156979,  1.25864397],
              [-0.13841039, -1.52834029,  0.64229365],
              [-1.60947559, -0.49054086,  0.08816671],
              [ 1.76189114, -0.69154256,  0.35327674]])
```

In Listing 5-2, you are ignoring rows b, d, and g. That's why you got NaN, which means non-numeric values. Pandas provides the `isnull()` and `notnull()` functions to detect the missing values in a data set. A Boolean value is returned when NaN has been detected; otherwise, False is returned, as shown in Listing 5-3.

Listing 5-3. Checking Null Cases

```
In [48]: print (dataset['stock1'].isnull())

a    False
b     True
c    False
d     True
e    False
f    False
g     True
h    False
Name: stock1, dtype: bool
```

Handling the Missing Values

There are various techniques that can be used to handle missing values.

- You can replace NaN with a scalar value.

  Listing 5-4 replaces all NaN cases with 0 values.

Listing 5-4. Replacing NaN with a Scalar Value

```
In [49]: print (dataset)
         dataset.fillna(0)

             stock1    stock2    stock3
         a -0.716435  0.646375  0.403254
         b       NaN       NaN       NaN
         c  0.923383 -0.354701 -0.594661
         d       NaN       NaN       NaN
         e  1.039185  0.984489  0.902545
         f -0.398857 -0.205501 -1.859085
         g       NaN       NaN       NaN
         h  0.228843  0.049838  0.400659

Out[31]:
```

	stock1	stock2	stock3
a	-0.716435	0.646375	0.403254
b	0.000000	0.000000	0.000000
c	0.923383	-0.354701	-0.594661
d	0.000000	0.000000	0.000000
e	1.039185	0.984489	0.902545
f	-0.398857	-0.205501	-1.859085
g	0.000000	0.000000	0.000000
h	0.228843	0.049838	0.400659

- You can fill NaN cases forward and backward.

  Another technique to handle missing values is to fill them forward using pad/fill or fill them backward using bfill/backfill methods. In Listing 5-5, the values of row a are replicating the missing values in row b.

Listing 5-5. Filling In Missing Values Forward

```
In [50]: # Fill missing values forward
         print (dataset)
         dataset.fillna(method='pad')
```

```
     stock1    stock2    stock3
a  0.512490  2.038219 -2.590846
b       NaN       NaN       NaN
c -1.187903 -0.301327  1.388822
d       NaN       NaN       NaN
e -0.892797  0.870075 -1.324887
f  1.227542  0.936045 -0.776875
g       NaN       NaN       NaN
h -1.570058 -0.363290  1.292037
```

Out[35]:

	stock1	stock2	stock3
a	0.512490	2.038219	-2.590846
b	0.512490	2.038219	-2.590846
c	-1.187903	-0.301327	1.388822
d	-1.187903	-0.301327	1.388822
e	-0.892797	0.870075	-1.324887
f	1.227542	0.936045	-0.776875
g	1.227542	0.936045	-0.776875
h	-1.570058	-0.363290	1.292037

- You can drop the missing values.

  Another technique is to exclude all the rows with NaN values. The Pandas `dropna()` function can be used to drop entire rows from the data set. As you can see in Listing 5-6, rows `b`, `d`, and `g` are removed entirely from the data set.

Listing 5-6. Dropping All NaN Rows

```
In [51]: print (dataset)
         dataset.dropna()

           stock1    stock2    stock3
a  0.884239  0.228564 -0.484426
b       NaN       NaN       NaN
c -0.274077  0.678091 -0.355736
d       NaN       NaN       NaN
e -1.937147  1.220786  0.243400
f -2.230833  0.183692  0.957954
g       NaN       NaN       NaN
h -0.984818  0.198828 -1.119425

Out[37]:
```

	stock1	stock2	stock3
a	0.884239	0.228564	-0.484426
c	-0.274077	0.678091	-0.355736
e	-1.937147	1.220786	0.243400
f	-2.230833	0.183692	0.957954
h	-0.984818	0.198828	-1.119425

- You can replace the missing (or generic) values.

  The `replace()` method can be used to replace a specific value in a data set with another given value. In addition, it can be used to replace NaN cases, as shown in Listing 5-7.

Listing 5-7. Using the replace() Function

```
In [52]: print (dataset)
         dataset.replace(np.nan, 0 )

             stock1    stock2    stock3
         a  0.830097 -0.149682 -1.532897
         b       NaN       NaN       NaN
         c -0.006940  0.750294 -0.772074
         d       NaN       NaN       NaN
         e -1.347131 -0.644828  0.465200
         f -0.853575  1.852128 -0.451999
         g       NaN       NaN       NaN
         h -0.308116  0.748715 -0.034594

Out[44]:
```

	stock1	stock2	stock3
a	0.830097	-0.149682	-1.532897
b	0.000000	0.000000	0.000000
c	-0.006940	0.750294	-0.772074
d	0.000000	0.000000	0.000000
e	-1.347131	-0.644828	0.465200
f	-0.853575	1.852128	-0.451999
g	0.000000	0.000000	0.000000
h	-0.308116	0.748715	-0.034594

Reading and Cleaning CSV Data

In this section, you will read data from a comma-separated values (CSV) file. The CSV sales file format shown in Figure 5-1 will be used to demonstrate the data cleaning process.

SALES_ID	SALES_BY_	JANUARY	FEBRUARY	MARCH	APRIL	MAY	JUNE	JULY	AUGUST	SEPTEMBE	OCTOBER	NOVEMBE	DECEMBER
1	AUH	3,469.00	n.a.	not avilab	3,642.00	5,803.00	5,662.00	1,896.00	2,293.00	2,583.00	5,233.00	4,421.00	4,071.00
1	SHJ	5,840.00	5,270.00	4,114.00	5,605.00	4,387.00	5,026.00	4,055.00	2,782.00	4,578.00	4,993.00	2,859.00	4,853.00
1	-1	2,967.00	2,425.00	5,353.00	n.a.	5,027.00	4,078.00	3,858.00	1,927.00	3,527.00	4,179.00	1,571.00	5,551.00
2	AUH	1,328.00	-1	1,574.00	2,343.00	3,826.00	4,932.00	1,710.00	3,221.00	3,381.00	1,313.00	1,765.00	1,214.00
3	SHJ	2,473.00	1,421.00	3,606.00	1,314.00	1,413.00	2,091.00	3,270.00	3,346.00	2,080.00	1,539.00	2,630.00	1,697.00
3	not avilab	n.a.	956	1,297.00	1,984.00	2,744.00	5,793.00	2,261.00	5,607.00	2,437.00	4,328.00	3,317.00	5,390.00
3	AUH	2,634.00	2,143.00	3,698.00	5,767.00	2,782.00	4,444.00	5,036.00	4,805.00	5,792.00	5,256.00	4,096.00	3,170.00
4	AJM	4,673.00	1,322.00	2,615.00	3,423.00	5,694.00	3,544.00	2,093.00	2,676.00	5,979.00	5,481.00	4,786.00	1,637.00
4	RAK	4,862.00	947	3,400.00	3,913.00	5,654.00	2,504.00	1,922.00	1,577.00	2,766.00	2,318.00	5,906.00	5,477.00
4	RAK	5,829.00	1,802.00	2,345.00	5,131.00	3,662.00	1,323.00	3,214.00	1,150.00	1,191.00	1,328.00	3,766.00	2,050.00
4	-1	4,870.00	4,645.00	5,521.00	4,824.00	3,899.00	1,118.00	3,665.00	5,241.00	2,433.00	5,215.00	1,858.00	4,667.00
5	FUJ	2,152.00	3,064.00	3,981.00	2,958.00	3,794.00	5,453.00	3,324.00	n.a.	1,516.00	5,640.00	1,301.00	5,513.00
5	AJM	5,584.00	1,103.00	2,838.00	3,520.00	2,250.00	1,435.00	3,035.00	4,572.00	5,450.00	2,374.00	1,549.00	4,885.00
5	DXB	4,258.00	1,397.00	3,824.00	4,715.00	3,150.00	5,572.00	968	1,014.00	2,784.00	5,913.00	969	4,153.00
5	SHJ	1,738.00	5,212.00	4,647.00	3,637.00	4,616.00	4,651.00	2,298.00	3,097.00	4,239.00	3,001.00	2,457.00	975
5	RAK	5,847.00	5,635.00	2,110.00	944	2,597.00	2,381.00	2,094.00	5,849.00	2,393.00	3,246.00	5,784.00	1,486.00
6	DXB	2,223.00	2,054.00	3,339.00	4,368.00	4,539.00	5,852.00	2,304.00	1,841.00	4,616.00	4,180.00	2,503.00	4,211.00
7	FUJ	5,564.00	1,972.00	3,522.00	2,779.00	1,371.00	5,419.00	1,398.00	4,277.00	5,467.00	4,831.00	1,507.00	3,749.00
7	FUJ	1,119.00	1,666.00	3,257.00	1,188.00	5,301.00	1,362.00	3,452.00	912	4,756.00	962	3,618.00	1,351.00
7	SHJ	1,723.00	1,196.00	2,044.00	925	2,973.00	1,581.00	1,812.00	4,429.00	1,230.00	5,813.00	2,119.00	3,105.00
7	RAK	972	1,938.00	5,041.00	2,119.00	3,172.00	2,327.00	2,104.00	4,796.00	2,454.00	1,500.00	4,799.00	2,987.00
8	DXB	3,490.00	not avilab	1,317.00	2,351.00	1,864.00	n.a.	4,161.00	1,131.00	2,173.00	4,005.00	5,238.00	1,396.00

***Figure 5-1.** Sales data in CSV format*

You can use the Pandas library to read a file and display the first five records. An autogenerated index has been generated by Python starting with 0, as shown in Listing 5-8.

Listing 5-8. Reading a CSV File and Displaying the First Five Records

```
In [53]: import pandas as pd
         sales = pd.read_csv("Sales.csv")
         print ("\n\n<<<<<<< First 5 records <<<<<<\n\n" )
         print (sales.head())
```

```
<<<<<<< First  5 records <<<<<<<

   SALES_ID SALES_BY_REGION   JANUARY  FEBRUARY              MARCH     APRIL  \
0         1             AUH  3,469.00      n.a.  not avilable  3,642.00
1         1             SHJ  5,840.00  5,270.00       4,114.00  5,605.00
2         1              -1  2,967.00  2,425.00       5,353.00      n.a.
3         2             AUH  1,328.00        -1       1,574.00  2,343.00
4         3             SHJ  2,473.00  1,421.00       3,606.00  1,314.00

        MAY      JUNE      JULY    AUGUST SEPTEMBER   OCTOBER  NOVEMBER  \
0  5,803.00  5,662.00  1,896.00  2,293.00  2,583.00  5,233.00  4,421.00
1  4,387.00  5,026.00  4,055.00  2,782.00  4,578.00  4,993.00  2,859.00
2  5,027.00  4,078.00  3,858.00  1,927.00  3,527.00  4,179.00  1,571.00
3  3,826.00  4,932.00  1,710.00  3,221.00  3,381.00  1,313.00  1,765.00
4  1,413.00  2,091.00  3,270.00  3,346.00  2,080.00  1,539.00  2,630.00

   DECEMBER
0  4,071.00
1  4,853.00
2  5,551.00
3  1,214.00
4  1,697.00
```

You can display the last five records using the `tail()` method.

```
In [54]:  print (sales.tail())
```

`pd.read_csv()` is used to read the entire CSV file; sometimes you need to read only a few records to reduce memory usage, though. In that case, you can use the `nrows` attribute to control the number of rows you want to read.

```
In [55]:   import pandas as pd
                salesNrows = pd.read_csv("Sales.csv", nrows=4)
                salesNrows
```

Similarly, you can read specific columns using a column index or label. Listing 5-9 reads columns 0, 1, and 6 using the `usecols` attribute and then uses the column labels instead of the column indices.

Listing 5-9. Renaming Column Labels

```
In [58]: salesNrows = pd.read_csv("Sales.csv", nrows=4,
usecols=[0, 1, 6])
         salesNrows
```

Out[10]:

	SALES_ID	SALES_BY_REGION	MAY
0	1	AUH	5,803.00
1	1	SHJ	4,387.00
2	1	-1	5,027.00
3	2	AUH	3,826.00

```
In [60]: salesNrows = pd.read_csv("Sales.csv", nrows=4,
usecols=['SALES_ID' , 'SALES_BY_REGION', 'FEBRUARY', 'MARCH'])
   salesNrows
```

Out[11]:

	SALES_ID	SALES_BY_REGION	FEBRUARY	MARCH
0	1	AUH	n.a.	not avilable
1	1	SHJ	5,270.00	4,114.00
2	1	-1	2,425.00	5,353.00
3	2	AUH	-1	1,574.00

In Listing 5-10, the `.rename()` method is used to change data set column labels (e.g., SALES_ID changed to ID). In addition, you set `inplace=True` to commit these changes to the original data set, not to a copy of it.

Listing 5-10. Renaming Column Labels

```
In [56]: salesNrows.rename(columns={"SALES_ID":'ID',"SALES_BY_
REGION":'REGION'}, inplace=True)
       salesNrows
```

Out[7]:

	ID	REGION	JANUARY	FEBRUARY	MARCH	APRIL	MAY	JUNE	JULY	AUGUST	SEPTEMBER	OCTOBER	NOVEMBER	DECEMBER
0	1	AUH	3,469.00	n.a.	not avilable	3,642.00	5,803.00	5,662.00	1,896.00	2,293.00	2,583.00	5,233.00	4,421.00	4,071.00
1	1	SHJ	5,840.00	5,270.00	4,114.00	5,605.00	4,387.00	5,026.00	4,055.00	2,782.00	4,578.00	4,993.00	2,859.00	4,853.00
2	1	-1	2,967.00	2,425.00	5,353.00	n.a.	5,027.00	4,078.00	3,858.00	1,927.00	3,527.00	4,179.00	1,571.00	5,551.00
3	2	AUH	1,328.00	-1	1,574.00	2,343.00	3,826.00	4,932.00	1,710.00	3,221.00	3,381.00	1,313.00	1,765.00	1,214.00

You can find the unique values in your data set variables; you just refer to each column as a variable or pattern that can be used for further processing. See Listing 5-11.

Listing 5-11. Finding Unique Values in Columns

```
In [57]: print (len(salesNrows['JANUARY'].unique()))
         print (len(salesNrows['REGION'].unique()))
         print (salesNrows['JANUARY'].unique())
```

```
4
3
['3,469.00' '5,840.00' '2,967.00' '1,328.00']
```

To get precise data, you can replace all values that are anomalies with NaN for further processing. For example, as shown in Listing 5-12, you can use `na_values =["n.a.", "not avilable", -1]` to generate NaN cases while you are reading the CSV file.

Listing 5-12. Automatically Replacing Matched Cases with NaN

```
In [61]: import pandas as pd
         sales = pd.read_csv("Sales.csv", nrows=7, na_values
         =["n.a.", "not avilable"])
         mydata = sales.head(7)
         mydata
```

Out[53]:

	SALES_ID	SALES_BY_REGION	JANUARY	FEBRUARY	MARCH	APRIL	MAY	JUNE	JULY	AUGUST	SEPTEMBER	OCTOBER	NOVEMBER	DECEME
0	1	AUH	3,469.00	NaN	NaN	3,642.00	5,803.00	5,662.00	1,896.00	2,293.00	2,583.00	5,233.00	4,421.00	4,071
1	1	SHJ	5,840.00	5,270.00	4,114.00	5,605.00	4,387.00	5,026.00	4,055.00	2,782.00	4,578.00	4,993.00	2,859.00	4,853
2	1	-1	2,967.00	2,425.00	5,353.00	NaN	5,027.00	4,078.00	3,858.00	1,927.00	3,527.00	4,179.00	1,571.00	5,551
3	2	AUH	1,328.00	-1	1,574.00	2,343.00	3,826.00	4,932.00	1,710.00	3,221.00	3,381.00	1,313.00	1,765.00	1,214
4	3	SHJ	2,473.00	1,421.00	3,606.00	1,314.00	1,413.00	2,091.00	3,270.00	3,346.00	2,080.00	1,539.00	2,630.00	1,697
5	3	NaN	NaN	956	1,297.00	1,984.00	2,744.00	5,793.00	2,261.00	5,607.00	2,437.00	4,328.00	3,317.00	5,390
6	3	AUH	2,634.00	2,143.00	3,698.00	5,767.00	2,782.00	4,444.00	5,036.00	4,805.00	5,792.00	5,256.00	4,096.00	3,170

```
In [62]: import pandas as pd
         sales = pd.read_csv("Sales.csv", nrows=7, na_values
         =["n.a.", "not avilable", -1])
         mydata = sales.head(7)
         mydata
```

Out[54]:

	SALES_ID	SALES_BY_REGION	JANUARY	FEBRUARY	MARCH	APRIL	MAY	JUNE	JULY	AUGUST	SEPTEMBER	OCTOBER	NOVEMBER	DECEME
0	1	AUH	3,469.00	NaN	NaN	3,642.00	5,803.00	5,662.00	1,896.00	2,293.00	2,583.00	5,233.00	4,421.00	4,071
1	1	SHJ	5,840.00	5,270.00	4,114.00	5,605.00	4,387.00	5,026.00	4,055.00	2,782.00	4,578.00	4,993.00	2,859.00	4,853
2	1	NaN	2,967.00	2,425.00	5,353.00	NaN	5,027.00	4,078.00	3,858.00	1,927.00	3,527.00	4,179.00	1,571.00	5,551
3	2	AUH	1,328.00	NaN	1,574.00	2,343.00	3,826.00	4,932.00	1,710.00	3,221.00	3,381.00	1,313.00	1,765.00	1,214
4	3	SHJ	2,473.00	1,421.00	3,606.00	1,314.00	1,413.00	2,091.00	3,270.00	3,346.00	2,080.00	1,539.00	2,630.00	1,697
5	3	NaN	NaN	956	1,297.00	1,984.00	2,744.00	5,793.00	2,261.00	5,607.00	2,437.00	4,328.00	3,317.00	5,390
6	3	AUH	2,634.00	2,143.00	3,698.00	5,767.00	2,782.00	4,444.00	5,036.00	4,805.00	5,792.00	5,256.00	4,096.00	3,170

Since you have different patterns in a data set, you should be able to use different values for data cleaning and replacement. The following example is reading from the `sales.csv` file and storing the data into the sales data frame. All values listed in the `na_values` attribute are replaced with the NaN value. So, for the January column, all `["n.a.", "not available", -1]` values are converted into NaN.

```
In [25]: sales = pd.read_csv("Sales.csv", na_values = {
                     "SALES_BY_REGION": ["n.a.", "not avilabl”],
                     "JANUARY": ["n.a.", "not avilable", -1],
                     "FEBRUARY": ["n.a.", "not avilable", -1]})
         sales.head(20)
```

Another professional method to clean data, while you are loading it, is to define functions for data cleaning. In Listing 5-13, you define and call two functions: `CleanData_Sales()` to clean numerical values and reset all NaN values to 0 and `CleanData_REGION()` to clean string values and reset all NaN values to `Abu Dhabi`. Then you call these functions in the `converters` attribute.

Listing 5-13. Defining and Calling Functions for Data Cleaning

```
In [26]: def CleanData_Sales(cell):
             if (cell=="n.a." or cell=="-1" or cell=="not
             avilable"):
                 return 0
             return cell

         def CleanData_REGION(cell):
             if (cell=="n.a." or cell=="-1" or cell=="not
             avilable"):
                 return 'AbuDhabi'
             return cell

In [28]: sales = pd.read_csv("Sales.csv", nrows=7, converters={
                                 "SALES_BY_REGION": CleanData_REGION,
                                 "JANUARY": CleanData_Sales,
                                 "FEBRUARY": CleanData_Sales,
                                 "APRIL": CleanData_Sales,
                                 })
         sales.head(20)
```

Out[27]:

	SALES_ID	SALES_BY_REGION	JANUARY	FEBRUARY	MARCH	APRIL	MAY	JUNE	JULY	AUGUST	SEPTEMBER	OCTOBER	NOVEMBER	DECEME
0	1	AUH	3,469.00	0	not avilable	3,642.00	5,803.00	5,662.00	1,896.00	2,293.00	2,583.00	5,233.00	4,421.00	4,071
1	1	SHJ	5,840.00	5,270.00	4,114.00	5,605.00	4,387.00	5,026.00	4,055.00	2,782.00	4,578.00	4,993.00	2,859.00	4,853
2	1	AbuDhabi	2,967.00	2,425.00	5,353.00	0	5,027.00	4,078.00	3,858.00	1,927.00	3,527.00	4,179.00	1,571.00	5,551
3	2	AUH	1,328.00	0	1,574.00	2,343.00	3,826.00	4,932.00	1,710.00	3,221.00	3,381.00	1,313.00	1,765.00	1,214
4	3	SHJ	2,473.00	1,421.00	3,606.00	1,314.00	1,413.00	2,091.00	3,270.00	3,346.00	2,080.00	1,539.00	2,630.00	1,697
5	3	AbuDhabi	0	956	1,297.00	1,984.00	2,744.00	5,793.00	2,261.00	5,607.00	2,437.00	4,328.00	3,317.00	5,390
6	3	AUH	2,634.00	2,143.00	3,698.00	5,767.00	2,782.00	4,444.00	5,036.00	4,805.00	5,792.00	5,256.00	4,096.00	3,170

Merging and Integrating Data

Python provides the `merge()` method to merge different data sets together using a specific common pattern. Listing 5-14 reads two different data sets about export values in a different range of years but for the same countries.

Listing 5-14. Two Files of Export Sales

```
In [35]: import pandas as pd

         a = pd.read_csv("1. Export1_Columns.csv")
         b = pd.read_csv("1. Export2_Columns.csv")
```

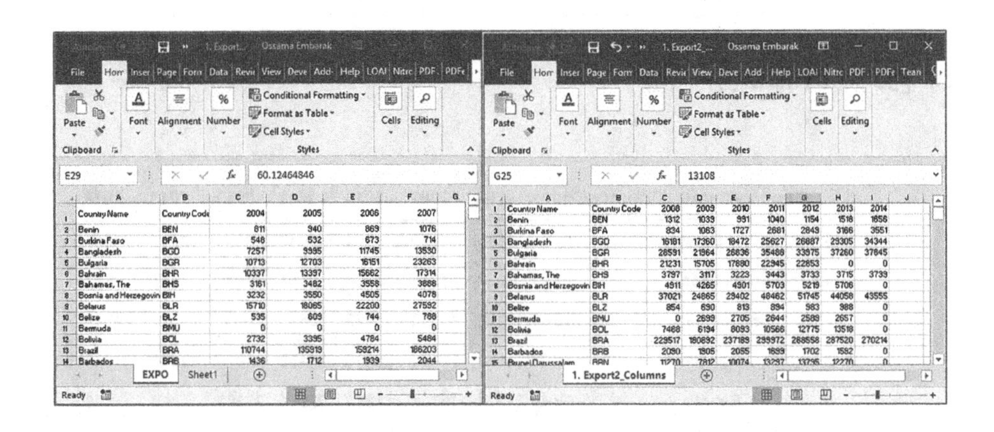

Suppose that you want to drop specific years from this study such as 2009, 2012, 2013, and 2014. Listing 5-15 and Listing 5-16 demonstrate different methods that are used to drop these columns.

Listing 5-15. Loading Two Different Data Sets with One Common Attribute

```
In [35]: import pandas as pd
         a = pd.read_csv("1. Export1_Columns.csv")
         b = pd.read_csv("1. Export2_Columns.csv")

In [31]: a.head()
```

Out[62]:

	Country Name	Country Code	2004	2005	2006	2007
0	Benin	BEN	811	940	869	1076
1	Burkina Faso	BFA	548	532	673	714
2	Bangladesh	BGD	7257	9995	11745	13530
3	Bulgaria	BGR	10713	12703	16151	23263
4	Bahrain	BHR	10337	13397	15662	17314

```
In [30]: b.head()
```

Out[61]:

	Country Name	Country Code	2008	2009	2010	2011	2012	2013	2014
0	Benin	BEN	1312	1039	991	1040	1154	1518	1656
1	Burkina Faso	BFA	834	1063	1727	2681	2849	3166	3551
2	Bangladesh	BGD	16181	17360	18472	25627	26887	29305	34344
3	Bulgaria	BGR	28591	21964	26836	35488	33975	37260	37845
4	Bahrain	BHR	21231	15705	17880	22945	22853	0	0

Listing 5-16. Dropping Columns 2009, 2012, 2013, and 2014

```
In [32]: b.drop('2014', axis=1, inplace=True)
       columns = ['2013', '2012']
         b.drop(columns, inplace=True, axis=1)
         b.head()
```

Out[68]:

	Country Name	Country Code	2008	2010	2011
0	Benin	BEN	1312	991	1040
1	Burkina Faso	BFA	834	1727	2681
2	Bangladesh	BGD	16181	18472	25627
3	Bulgaria	BGR	28591	26836	35488
4	Bahrain	BHR	21231	17880	22945

Python's `.merge()` method can used to merge data sets; you can specify the merging variables, or you can let Python find the matching variables and implement the merging, as shown in Listing 5-17.

Listing 5-17. Merging Two Data Sets

```
In [102]: mergedDataSet = a.merge(b, on="Country Name")
          mergedDataSet.head()
```

Merge two datasets using column labeled County Code_x and County Code_y as shown below.

Out[69]:

	Country Name	Country Code_x	2004	2005	2006	2007	Country Code_y	2008	2010	2011
0	Benin	BEN	811	940	869	1076	BEN	1312	991	1040
1	Burkina Faso	BFA	548	532	673	714	BFA	834	1727	2681
2	Bangladesh	BGD	7257	9995	11745	13530	BGD	16181	18472	25627
3	Bulgaria	BGR	10713	12703	16151	23263	BGR	28591	26836	35488
4	Bahrain	BHR	10337	13397	15662	17314	BHR	21231	17880	22945

```
In [103]:  dataX = a.merge(b)
           dataX.head()
```

Out[70]:

	Country Name	Country Code	2004	2005	2006	2007	2008	2010	2011
0	Benin	BEN	811	940	869	1076	1312	991	1040
1	Burkina Faso	BFA	548	532	673	714	834	1727	2681
2	Bangladesh	BGD	7257	9995	11745	13530	16181	18472	25627
3	Bulgaria	BGR	10713	12703	16151	23263	28591	26836	35488
4	Bahrain	BHR	10337	13397	15662	17314	21231	17880	22945

You can merge two data sets using Index via Rows Union operation, as indicated in Listing 5-18, where the `.concat()` method is used to merge `Data1` and `Data2` over axis 0. This is a row-wise operation.

Listing 5-18. Row Union of Two Data Sets

```
In [71]: Data1 = a.head()
         Data1=Data1.reset_index()
         Data1
```

Out[71]:

	index	Country Name	Country Code	2004	2005	2006	2007
0	0	Benin	BEN	811	940	869	1076
1	1	Burkina Faso	BFA	548	532	673	714
2	2	Bangladesh	BGD	7257	9995	11745	13530
3	3	Bulgaria	BGR	10713	12703	16151	23263
4	4	Bahrain	BHR	10337	13397	15662	17314

```
In [72]: Data2 = a.tail()
         Data2=Data2.reset_index()
         Data2
```

Out[72]:

	index	Country Name	Country Code	2004	2005	2006	2007
0	228	Yemen, Rep.	YEM	5048	6852	7873	0
1	229	South Africa	ZAF	58216	68172	79519	93339
2	230	Congo, Dem. Rep.	COD	2341	2442	2765	6540
3	231	Zambia	ZMB	2087	2550	4158	4722
4	232	Zimbabwe	ZWE	2001	1931	1957	2000

```
In [78]: # stack the DataFrames on top of each othe
         VerticalStack = pd.concat((Data1, Data2), axis=0)
         VerticalStack
```

Out[73]:

	Index	Country Name	Country Code	2004	2005	2006	2007
0	0	Benin	BEN	811	940	869	1076
1	1	Burkina Faso	BFA	548	532	673	714
2	2	Bangladesh	BGD	7257	9995	11745	13530
3	3	Bulgaria	BGR	10713	12703	16151	23263
4	4	Bahrain	BHR	10337	13397	15662	17314
0	228	Yemen, Rep.	YEM	5048	6852	7873	0
1	229	South Africa	ZAF	58216	68172	79519	93339
2	230	Congo, Dem. Rep.	COD	2341	2442	2765	6540
3	231	Zambia	ZMB	2087	2550	4158	4722
4	232	Zimbabwe	ZWE	2001	1931	1957	2000

Reading Data from the JSON Format

The Pandas library can read JSON files using the `read_json` function directly from the cloud or from a hard disk. Listing 5-19 demonstrates how to create JSON data and load it in JSON format and then iterate or manipulate the data. The JSON format is similar to a dictionary structure where you have a key-value pair, but in JSON, you can have subattributes with inner values, similar to `email` in the first example, and its subattribute `hide` with the value `NO`.

Listing 5-19. Creating and Manipulating JSON Data

```
In [73]: import json data = '''{
         "name" : "Ossama",
         "phone" : { "type" : "intl", "number" : "+971 50 244
         5467"},
         "email" : {"hide" : "No" }
         }'''
```

```
        info = json.loads(data)
        print ('Name:',info["name"])
        print ('Hide:',info["email"]["hide"])
        Name: Ossama
        Hide: No

In [74]: input = '''[
        { "id" : "001", "x" : "5", "name" : "Ossama"} ,
        { "id" : "009","x" : "10","name" : "Omar" }
        ]'''
        info = json.loads(input) print ('User count:',
        len(info)) for item in info:
            print ('\nName', item['name'])
            print ('Id', item['id'])
            print ('Attribute', item['x'])
        User count: 2

        Name Ossama
        Id 001
        Attribute 5

        Name Omar
        Id 009
        Attribute 10
```

You can directly read JSON data from an online resource, as shown in Listing 5-20 and Listing 5-21.

Listing 5-20. JSON Sample Data

```
url=' http://python-data.dr-chuck.net/comments_244984.json'
print ('Retrieving', url)
uh = urllib.urlopen(url)
data = uh.read()
```

```
▼ comments:
  ▼ 0:
      name:    "Abaan"
      count:   98
  ▼ 1:
      name:    "Ashna"
      count:   95
  ▼ 2:
      name:    "Dante"
      count:   94
  ▼ 3:
      name:    "Isabel"
      count:   93
  ▼ 4:
      name:    "Fearne"
      count:   92
```


Listing 5-21. Loading a JSON File

```
In [101]: import json
            with open('comments.json') as json_data:
              JSONdta = json.load(json_data)

              print(JSONdta)
```


```
{'note': 'This file contains the actual data for your assignment', 'comments': [{'name': 'Abaan', 'count': 98}, {'nam
e': 'Ashna', 'count': 95}, {'name': 'Dante', 'count': 94}, {'name': 'Isabel', 'count': 93}, {'name': 'Fearne', 'count
': 92}, {'name': 'Kriss', 'count': 91}, {'name': 'Janani', 'count': 87}, {'name': 'Karhys', 'count': 85}, {'name': 'M
egg', 'count': 84}, {'name': 'Luisa', 'count': 83}, {'name': 'Thorben', 'count': 79}, {'name': 'Kaelan', 'count': 77}
, {'name': 'Ceirin', 'count': 75}, {'name': 'Lileidh', 'count': 70}, {'name': 'Angelika', 'count': 70}, {'name': 'Ame
lka', 'count': 69}, {'name': 'Justin', 'count': 69}, {'name': 'Muneeb', 'count': 68}, {'name': 'Antoine', 'count': 64
}, {'name': 'Ivar', 'count': 61}, {'name': 'Kaid', 'count': 60}, {'name': 'Dakotah', 'count': 58}, {'name': 'Nadeem',
'count': 58}, {'name': 'Marybeth', 'count': 55}, {'name': 'Ashlyn', 'count': 55}, {'name': 'Kaydin', 'count': 50}, {'
name': 'Obieluem', 'count': 48}, {'name': 'Cairn', 'count': 46}, {'name': 'Ala', 'count': 45}, {'name': 'Vithujan', '
count': 38}, {'name': 'Ivory', 'count': 34}, {'name': 'Rosalyn', 'count': 33}, {'name': 'Kaywan', 'count': 32}, {'nam
e': 'Pedro', 'count': 31}, {'name': 'Bharath', 'count': 30}, {'name': 'Eshaal', 'count': 29}, {'name': 'Aliya', 'coun
t': 28}, {'name': 'Sephiroth', 'count': 27}, {'name': 'Minah', 'count': 25}, {'name': 'Murdo', 'count': 22}, {'name':
'Ata', 'count': 21}, {'name': 'Remonae', 'count': 17}, {'name': 'Muskaan', 'count': 17}, {'name': 'Lottie', 'count':
17}, {'name': 'Glane', 'count': 9}, {'name': 'Dineo', 'count': 6}, {'name': 'Zoe', 'count': 5}, {'name': 'Raul', 'cou
nt': 4}, {'name': 'Tammylee', 'count': 2}, {'name': 'Morna', 'count': 1}]}
```


You can access JSON data and make further operations on the extracted data. For instance, you can calculate the total number of all users, find the average value of all counts, and more, as shown in Listing 5-22.

Listing 5-22. Accessing JSON Data

```
In [102]:sumv=0
         counter=0
         for i in range(len(JSONdta["comments"])):
             counter+=1
             Name = JSONdta["comments"][i]["name"]
             Count = JSONdta["comments"][i]["count"]
             sumv+=int(Count)
         print (Name," ", Count)
         print ("\nCount: ", counter)
         print ("Sum: ", sumv)
```

The following is a sample of extracted data from the JSON file and the calculated total number of all users:

```
Murdo   22
Ata   21
Remonae   17
Muskaan   17
Lottie   17
Giane   9
Dineo   6
Zoe   5
Raul   4
Tammylee   2
Morna   1

Count:  50
Sum:  2507
```

Reading Data from the HTML Format

You can read online HTML files, but you should install and use the Beautiful Soup package to do so. Listing 5-23 shows how to make a request to a URL to be loaded into the Python environment. Then you use the

HTML parser parameter to read the entire HTML file. You can also extract values stored with HTML tags.

Listing 5-23. Reading and Parsing an HTML File

```
In [104]:import urllib from bs4
         import BeautifulSoup
         response = urllib.request.urlopen('http://python-data.
         dr-chuck.net/known_by_Rona.html'
         html_doc = response.read()
         soup = BeautifulSoup(html_doc, 'html.parser')
         print(html_doc[:700])
         print("\n")
         print (soup.title)
         print(soup.title.string)
         print(soup.a.string)
```

```
b'<html>\n<head>\n<title>People that Rona knows</title>\n<style>\n.overlay{\n    opacity:0.99;\n    background-color:
#eee;\n    position:fixed;\n    width:100%;\n    height:100%;\n    top:0px;\n    left:0px;\n    z-index:1000;\n}\n</s
tyle>\n</head>\n<body>\n<h1>People that Rona knows</h1>\n<div class="overlay" id="overlay" style="display:none" >\n<c
enter>\n<h2>\nThis screen randomly changes the height between list items and vanishes \nafter a while to make sure th
at you retrieve and process the data\nin a Python program rather than simply counting down pressing links, and \ndoin
g the assignment without writing a Python program :).\nThe names are in the same order in the HTML even though they \
nshift around on the scree'


<title>People that Rona knows</title>
People that Rona knows
Konar
```

```
In [103]: import urllib.request
          with urllib.request.urlopen("http://python-data.dr-
          chuck.net/known_by_Rona.html") as url:
                  strhtml = url.read()
          #I'm guessing this would output the html source code?
          print(strhtml[:700])
```

```
b'<html>\n<head>\n<title>People that Rona knows</title>\n<style>\n.overlay{\n    opacity:0.99;\n    background-color:
#eee;\n    position:fixed;\n    width:100%;\n    height:100%;\n    top:0px;\n    left:0px;\n    z-index:1000;\n}\n</s
tyle>\n</head>\n<body>\n<h1>People that Rona knows</h1>\n<div class="overlay" id="overlay" style="display:none" >\n<c
enter>\n<h2>\nThis screen randomly changes the height between list items and vanishes \nafter a while to make sure th
at you retrieve and process the data\nin a Python program rather than simply counting down pressing links, and \ndoin
g the assignment without writing a Python program :).\nThe names are in the same order in the HTML even though they \
```

You can also load HTML and use the Beautiful Soup package to parse HTML tags and display the first ten anchor tags, as shown in Listing 5-24.

Listing 5-24. Parsing HTML Tags

```
In [107]: import urllib from bs4
import BeautifulSoup
          response = urllib.request.urlopen('http://python-
          data.dr chuck.net/known_by_Rona.html' html_doc =
          response.read()
          print (html_doc[:300])
          soup = BeautifulSoup(html_doc, 'html.parser')
          print ("\n") counter=0
          for link in soup.findAll("a"):
              print(link.get("href"))
              if counter<10: counter+=1
                   continue
              else:
                   break
```

```
b'<html>\n<head>\n<title>People that Rona knows</title>\n<style>\n.overlay{\n    opacity:0.99;\n    background-color:#eee;\
n    position:fixed;\n    width:100%;\n    height:100%;\n    top:0px;\n    left:0px;\n    z-index:1000;\n}\n</style>\n</hea
d>\n<body>\n<h1>People that Rona knows</h1>\n<div class="overlay" id="over'


http://python-data.dr-chuck.net/known_by_Konar.html
http://python-data.dr-chuck.net/known_by_Mohamad.html
http://python-data.dr-chuck.net/known_by_Keyra.html
http://python-data.dr-chuck.net/known_by_Jaxson.html
http://python-data.dr-chuck.net/known_by_Jordyn.html
http://python-data.dr-chuck.net/known_by_Cairn.html
http://python-data.dr-chuck.net/known_by_Bodhan.html
http://python-data.dr-chuck.net/known_by_Arianna.html
http://python-data.dr-chuck.net/known_by_Kiarrah.html
http://python-data.dr-chuck.net/known_by_Alannah.html
http://python-data.dr-chuck.net/known_by_Keira.html
```

Let's create an `html` variable that maintains some web page content and read it using Beautiful Soup, as shown in Listing 5-25.

Listing 5-25. Reading HTML Using Beautiful Soup

```
In [108]: htmldata="""<html>
            <head>
            <title>
              The Dormouse's story
            </title>
            </head>
            <body>
              <p class="title">
                <b>
                  The Dormouse's story
                </b>
              </p>
              <p class="story">
                Once upon a time there were three little
                sisters; and their names were
                <a class="sister" href="http://example.com/
                elsie" id="link1"> Elsie
                </a>
                ,
                <a class="sister" href="http://example.com/
                lacie" id="link2"> Lacie
                </a> and
                <a class="sister" href="http://example.com/
                tillie" id="link2"> Tillie
                </a>
                ; and they lived at the bottom of a well.
              </p>
```

```
        <p class="story"> ...
        </p>
      </body>
    </html>
"""
from bs4 import BeautifulSoup
soup = BeautifulSoup(htmldata, 'html.parser')
print(soup.prettify())
```

```
<html>
 <head>
  <title>
   The Dormouse's story
  </title>
 </head>
 <body>
  <p class="title">
   <b>
    The Dormouse's story
   </b>
  </p>
  <p class="story">
   Once upon a time there were three little sisters; and their names were
   <a class="sister" href="http://example.com/elsie" id="link1">
    Elsie
   </a>
   ,
   <a class="sister" href="http://example.com/lacie" id="link2">
    Lacie
   </a>
   and
   <a class="sister" href="http://example.com/tillie" id="link2">
    Tillie
   </a>
   ; and they lived at the bottom of a well.
  </p>
  <p class="story">
   ...
  </p>
 </body>
</html>
```

You can also use Beautiful Soup to extract data from HTML. You can extract data, tags, or all related data such as all hyperlinks in the parsed HTML content, as shown in Listing 5-26.

Listing 5-26. Using Beautiful Soup to Extract Data from HTML

```
In [109]: soup.title
Out[109]: <title>
              The Dormouse's story
          </title>

In [110]: soup.title.name
Out[110]: 'title'

In [111]: soup.title.string
Out[111]: "\n      The Dormouse's story\n "

In [112]: soup.title.parent.name
Out[112]: 'head'

In [113]: soup.p
Out[113]: <p class="title">
            <b>
              The Dormouse's story
            </b>
          </p>

In [114]: soup.p['class']
Out[114]: ['title']

In [115]: soup.a
Out[115]: <a class="sister" href="http://example.com/elsie"
id="link1"> Elsie
          </a>
```

```
In [116]: soup.find_all('a')
Out[116]: [<a class="sister" href="http://example.com/elsie"
id="link1"> Elsie
          </a>, <a class="sister" href="http://example.com/
          lacie" id="link2"> Lacie
          </a>, <a class="sister" href="http://example.com/
          tillie" id="link2"> Tillie
          </a>]

In [117]: soup.find(id="link2")
Out[117]: <a class="sister" href="http://example.com/lacie"
id="link2"> Lacie
          </a>
```

It is possible to extract all the URLs found within a page's <a> tags, as shown in Listing 5-27.

Listing 5-27. Extracting All URLs in Web Page Content

```
In [118]: for link in soup.find_all('a'):
              print(link.get('href'))

http://example.com/elsie
http://example.com/lacie
http://example.com/tillie
```

Another common task is extracting all the text from a page and ignoring all the tags, as shown in Listing 5-28.

Listing 5-28. Extracting Only the Contents

```
In [119]: print(soup.get_text())

The Dormouse's story



 The Dormouse's story


Once upon a time there were three little sisters; and their names were

 Elsie

,

 Lacie

and

 Tillie

; and they lived at the bottom of a well.

...
```

Reading Data from the XML Format

Python provides the `xml.etree.ElementTree` (`ET`) module to implement a simple and efficient parsing of XML data. `ET` has two classes for this purpose: `ElementTree`, which represents the whole XML document as a tree, and `Element`, which represents a single node in this tree. Interactions with the whole document (reading and writing to/from files) are usually done on the `ElementTree` level. The interactions with a single XML element and its subelements are done on the `Element` level. In Listing 5-29, you are creating an XML container and reading it using `ET` for parsing purposes. Then you extract data from the container using the `find()` and `get()` methods, parsing through the generated tree.

Listing 5-29. Reading XML and Extracting Its Data

```
In [128]: xmldata = """
              <?xml version="1.0"?>
              <data>
                  <student
                      name="Omar">
                      <rank>2</rank>
                      <year>2017</year>
                      <GPA>3.5</GPA>
                      <concentration name="Networking"
                      Semester="7"/> </student>
                  <student name="Ali">
                      <rank>3</rank>
                      <year>2016</year>
                      <GPA>2.8</GPA>
                      <concentration name="Security"
                      Semester="6"/>
                  </student>

                  <student name="Osama">
                      <rank>1</rank>
                      <year>2018</year>
                      <GPA>3.7</GPA>
                      <concentration name="App Development"
                      Semester="8"/> </student>
              </data>
          """.strip()

In [129]:from xml.etree import ElementTree as ET stuff =
ET.fromstring(xmldata) lst = stuff.findall('student')

         print ('Students count:', len(lst)) for item in lst:
```

```
print ("\nName:", item.get("name"))
print ('concentration:', item.
find("concentration").get("name"))
print ('Rank:', item.find('rank').text)
print ('GPA:', item.find("GPA").text)
```

```
Students count: 3

Name: Omar
concentration: Networking
Rank: 2
GPA: 3.5

Name: Ali
concentration: Security
Rank: 3
GPA: 2.8

Name: Osama
concentration: App Development
Rank: 1
GPA: 3.7
```

Summary

This chapter covered data gathering and cleaning so that you can have reliable data for analysis. This list recaps what you studied in this chapter:

- How to apply cleaning techniques to handle missing values
- How to read CSV-formatted data offline and directly from the cloud

- How to merge and integrate data from different sources
- How to read and extract data from JSON, HTML, and XML formats

The next chapter will study how to explore and analyze data and much more.

Exercises and Answers

1. Write a Python script to read the data in an Excel file named `movies.xlsx` and save this data in a data frame called `mov`. Perform the following steps:

```
mov = pd.read_excel("movies.xlsx")
```

   a. Read the contents of the second sheet that is named 2000s in the Excel file (`movies.xlsx`) and store this content in a data frame called `Second_sheet`.

```
Second_sheet = pd.read_excel("movies.xlsx",sheetname = "2000s")
```

   b. Write the code needed to show the first seven rows from the data frame `Second_sheet` using an appropriate method.

```
Second_sheet.head(7)
```

   c. Write the code needed to show the last five rows using an appropriate method.

```
Second_sheet.tail()
```

d. Use a suitable command to show only one column that is named Budget.

```
Second_sheet["Budget"]
```

e. Use a suitable command to show the total rows in the first sheet that is called 2000s.

```
len(Second_sheet)
```

f. Use a suitable command to show the maximum value stored in the Budget column.

```
Second_sheet["Budget"].max()
```

g. Use a suitable command to show the minimum value stored in the Budget column.

```
Second_sheet["Budget"].min()
```

h. Write a single command to show the details (count, min, max, mean, std, 25%, 50%, 75%) about the column User Votes.

```
Second_sheet["User Votes"].describe()
```

i. Use a suitable conditional statement that stores the rows in which the country name is USA and the Duration value is less than 50 in a data frame named USA50. Show the values in data frame USA50.

```
USA50 = Second_sheet[(Second_sheet["Country"] == 'USA') & Second_sheet["Duration"] < 50]
USA50
```

j. Using a suitable command, create a calculated column named `Avg Reviews` in `Second_sheet` by adding Reviews by Users and Reviews by Critics and divide it by 2. Display the first five rows of the `Second_sheet` after creating the previous calculated column.

```
Second_sheet["Avg Reviews"] = (Second_sheet["Reviews by Users"] + Second_sheet["Reviews by Crtiics"] )/2
Second_sheet.head()
```

k. Using a suitable command, sort the `Country` values in ascending order (smallest to largest) and `Avg_reviews` in descending order (largest to smallest).

```
Second_sheet.sort_values(["Country","Avg Reviews"],ascending=[1,0])
```

l. Write a Python script to read the following HTML and extract and display only the content, ignoring the tag structure:

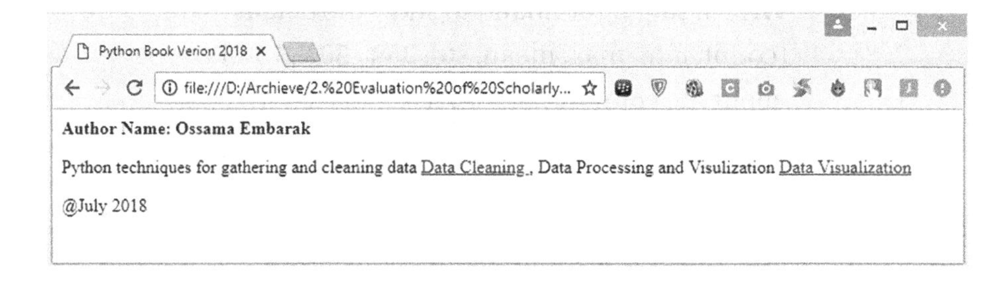

```
<html>
 <head>
  <title>
   Python Book Version 2018
  </title>
 </head>
 <body>
  <p class="title">
   <b>
    Author Name: Ossama Embarak
   </b>
  </p>
  <p class="story">
   Python techniques for gathering and cleaning data
   <a class="sister" href="https://leanpub.com/
   AgilePythonProgrammingAppliedForEveryone" id="link1">
    Data Cleaning
   </a>
   , Data Processing and Visualization
   <a class="sister" href="http://www.lulu.com/shop/ossama-
   embarak/agile-python-programming-applied-for-everyone/
   paperback/product-23694020.html" id="link2">
    Data Visualization
   </a>

  </p>
  <p class="story">
  @July 2018
  </p>
 </body>
</html>
```

Answer:

```
from bs4 import BeautifulSoup
soup = BeautifulSoup(htmldata, 'html.parser')
print(soup.prettify())
print(soup.get_text())
```

```
In [17]: htmldata="""<html>
           <head>
            <title>
             Python Book Verion 2018
            </title>
           </head>
           <body>
            <p class="title">
             <b>
              Author Name: Ossama Embarak
             </b>
            </p>
            <p class="story">
             Python techniques for gathering and cleaning data
             <a class="sister"

         href="https://leanpub.com/AgilePythonProgrammingAppliedForEveryone"

         id="link1">
              Data Cleaning
             </a>
             , Data Processing and Visulization
             <a class="sister" href="http://www.lulu.com/shop/ossama-

         embarak/agile-python-programming-applied-for-

         everyone/paperback/product-23694020.html" id="link2">
              Data Visualization
             </a>

            </p>
            <p class="story">
            @July 2018
            </p>
           </body>
          </html>
         """

         from bs4 import BeautifulSoup
         soup = BeautifulSoup(htmldata, 'html.parser')
         print(soup.prettify())
```

```
<html>
 <head>
  <title>
   Python Book Verion 2018
  </title>
 </head>
 <body>
  <p class="title">
   <b>
    Author Name: Ossama Embarak
   </b>
  </p>
  <p class="story">
   Python techniques for gathering and cleaning data
   <a class="sister" href="https://leanpub.com/AgilePythonProgrammingAppliedForEveryone" id="link1">
    Data Cleaning
   </a>
   , Data Processing and Visulisation
   <a class="sister" href="http://www.lulu.com/shop/ossama-

embarak/agile-python-programming-applied-for-

everyone/paperback/product-23694020.html" id="link2">
    Data Visualisation
   </a>
  </p>
  <p class="story">
   @July 2018
  </p>
 </body>
</html>
```

```
In [18]: print(soup.get_text())
```

```


    Python Book Verion 2018




     Author Name: Ossama Embarak


    Python techniques for gathering and cleaning data

     Data Cleaning

    , Data Processing and Visulisation

     Data Visualisation
```

CHAPTER 6

Data Exploring and Analysis

Nowadays, massive data is collected daily and distributed over various channels. This requires efficient and flexible data analysis tools. Python's open source Pandas library fills that gap and deals with three different data structures: series, data frames, and panels. A *series* is a one-dimensional data structure such as a dictionary, array, list, tuple, and so on. A *data frame* is a two-dimensional data structure with heterogeneous data types, i.e., tabular data. A *panel* refers to a three-dimensional data structure such as a three-dimensional array. It should be clear that the higher-dimensional data structure is a container of its lower-dimensional data structure. In other words, a panel is a container of a data frame, and a data frame is a container of a series.

Series Data Structures

As mentioned earlier, a series is a sequence of one-dimensional data such as a dictionary, list, array, tuple, and so on.

O. Embarak, *Data Analysis and Visualization Using Python*,
https://doi.org/10.1007/978-1-4842-4109-7_6

Creating a Series

Pandas provides a `Series()` method that is used to create a series structure. A serious structure of size `n` should have an index of length `n`. By default Pandas creates indices starting at 0 and ending with `n-1`. A Pandas series can be created using the constructor `pandas.Series (data, index, dtype, copy)` where `data` could be an array, constant, list, etc. The series `index` should be unique and hashable with length `n`, while `dtype` is a data type that could be explicitly declared or inferred from the received data. Listing 6-1 creates a series with a default index and with a set index.

Listing 6-1. Creating a Series

```
In [5]: import pandas as pd
        import numpy as np
        data = np.array(['O','S','S','A'])
        S1 = pd.Series(data) # without adding index
        S2 = pd.Series(data,index=[100,101,102,103]) # with
        adding index print (S1) print ("\n") print (S2)
        0    O
        1    S
        2    S
        3    A
        dtype: object

        100  O
        101  S
        102  S
        103  A
        dtype: object
```

```
In [40]:import pandas as pd
        import numpy as np
        my_series2 = np.random.randn(5, 10)
        print ("\nmy_series2\n", my_series2)
```

This is the output of creating a series of random values of 5 rows and 10 columns.

```
my_series2
 [[ 0.08590877  0.59702919 -1.29330859 -1.42021041 -0.09535271  0.09058623
  -1.14191133 -0.84699991  0.94028641  1.79400706]
 [ 0.50645411 -0.37674882 -1.16751734 -1.24061761  0.03981985  0.13478382
   0.76132521 -0.40671662 -0.7484758   0.30420489]
 [-0.66951224 -1.19373055  1.86446782  1.43047631 -0.06302096  0.49239499
  -0.48208329 -1.9805521  -0.73735706 -1.03152802]
 [-0.79181088  1.02769491 -1.27216885  0.20320462  0.19385809 -0.51614599
  -0.66898612 -0.60962025 -1.43724096 -0.22663712]
 [ 1.14193093 -0.8842498   0.22409272 -0.29599594  1.1917404   1.09016684
   1.87701454  1.08452103 -1.49587483 -0.31887386]]
```

As mentioned earlier, you can create a series from a dictionary; Listing 6-2 demonstrates how to create an index for a data series.

Listing 6-2. Creating an Indexed Series

```
In [6]: import pandas as pd
        import numpy as np
        data = {'X' : 0., 'Y' : 1., 'Z' : 2.}
        SERIES1 = pd.Series(data)
        print (SERIES1)
        X 0.0
        Y 1.0
        Z 2.0
        dtype: float64
In [7]: import pandas as pd
        import numpy as np
        data = {'X' : 0., 'Y' : 1., 'Z' : 2.}
        SERIES1 = pd.Series(data,index=['Y','Z','W','X'])
        print (SERIES1)
        Y 1.0
```

```
Z 2.0
W NaN
X 0.0
dtype: float64
```

If you can create series data from a scalar value as shown in Listing 6-3, then an index is mandatory, and the scalar value will be repeated to match the length of the given index.

Listing 6-3. Creating a Series Using a Scalar

```
In [9]: # Use sclara to create a series
        import pandas as pd
        import numpy as np
        Series1 = pd.Series(7, index=[0, 1, 2, 3, 4])
        print (Series1)
        0    7
        1    7
        2    7
        3    7
        4    7
        dtype: int64
```

Accessing Data from a Series with a Position

Like lists, you can access a series data via its index value. The examples in Listing 6-4 demonstrate different methods of accessing a series of data. The first example demonstrates retrieving a specific element with index 0. The second example retrieves indices 0, 1, and 2. The third example retrieves the last three elements since the starting index is -3 and moves backward to -2, -1. The fourth and fifth examples retrieve data using the series index labels.

Listing 6-4. Accessing a Data Series

```
In [18]: import pandas as pd
         Series1 = pd.Series([1,2,3,4,5],index =
                             ['a','b','c','d','e'])
         print ("Example 1:Retrieve the first element")
         print (Series1[0] )
         print ("\nExample 2:Retrieve the first three element")
         print (Series1[:3])
         print ("\nExample 3:Retrieve the last three element")
         print(Series1[-3:])
         print ("\nExample 4:Retrieve a single element")
         print (Series1['a'])
         print ("\nExample 5:Retrieve multiple elements")
         print (Series1[['a','c','d']])
```

```
Example 1:Retrieve the first element
1

Example 2:Retrieve the first three element
a    1
b    2
c    3
dtype: int64

Example 3:Retrieve the last three element
c    3
d    4
e    5
dtype: int64

Example 4:Retrieve a single element
1

Example 5:Retrieve multiple elements
a    1
c    3
d    4
dtype: int64
```

Exploring and Analyzing a Series

Numerous statistical methods can be applied directly on a data series. Listing 6-5 demonstrates the calculation of mean, max, min, and standard deviation of a data series. Also, the `.describe()` method can be used to give a data description, including quantiles.

Listing 6-5. Analyzing Series Data

```
In [10]: import pandas as pd
         import numpy as np
         my_series1 = pd.Series([5, 6, 7, 8, 9, 10])
         print ("my_series1\n", my_series1)
         print ("\n Series Analysis\n ")
         print ("Series mean value : ", my_series1.mean()) #
         find mean value in a series
         print ("Series max value : ",my_series1.max()) #
         find max value in a series
         print ("Series min value : ",my_series1.min()) #
         find min value in a series
         print ("Series standard deviation value : ",
         my_series1.std()) # find standard deviation
         my_series1
         0    5
         1    6
         2    7
         3    8
         4    9
         5    10
         dtype: int64
```

```
         Series Analysis

         Series mean value : 7.5
         Series max value : 10
         Series min value : 5
         Series standard deviation value : 1.8708286933869707

In [11]: my_series1.describe()
Out[11]: count     6.000000
         mean      7.500000
         std       1.870829
         min       5.000000
         25%       6.250000
         50%       7.500000
         75%       8.750000
         max      10.000000
         dtype: float64
```

If you copied by reference one series to another, then any changes to the series will adapt to the other one. After copying my_series1 to my_series_11, once you change the indices of my_series_11, it reflects back to my_series1, as shown in Listing 6-6.

Listing 6-6. Copying a Series to Another with a Reference

```
In [17]: my_series_11 = my_series1
         print (my_series1)
         my_series_11.index = ['A', 'B', 'C', 'D', 'E', 'F']
         print (my_series_11)
         print (my_series1)
         0    5
         1    6
         2    7
         3    8
```

```
4    9
5    10
dtype: int64
A    5
B    6
C    7
D    8
E    9
F    10
dtype: int64
A    5
B    6
C    7
D    8
E    9
F    10
dtype: int64
```

You can use the `.copy()` method to copy the data set without having a reference to the original series. See Listing 6-7.

Listing 6-7. Copying Series Values to Another

```
In [21]: my_series_11 = my_series1.copy()
         print (my_series1)
         my_series_11.index = ['A', 'B', 'C', 'D', 'E', 'F']
         print (my_series_11)
         print (my_series1)
         0    5
         1    6
         2    7
         3    8
```

```
4    9
5    10
dtype: int64
A    5
B    6
C    7
D    8
E    9
F    10
dtype: int64
0    5
1    6
2    7
3    8
4    9
5    10
dtype: int64
```

Operations on a Series

Numerous operations can be implemented on series data. You can check whether an index value is available in a series or not. Also, you can check all series elements against a specific condition, such as if the series value is less than 8 or not. In addition, you can perform math operations on series data directly or via a defined function, as shown in Listing 6-8.

Listing 6-8. Operations on Series

```
In [23]: 'F' in my_series_11
Out[23]: True

In [27]: temp = my_series_11 < 8
         temp
```

```
Out[27]: A     True
         B     True
         C     True
         D     False
         E     False
         F     False
         dtype: bool
         In [35]: len(my_series_11)

Out[35]: 6


In [28]: temp = my_series_11[my_series_11 < 8 ] * 2
         temp

Out[28]: A     10
         B     12
         C     14
         dtype: int64
```

Define a function to add two series and call the function, like this:

```
In [37]: def AddSeries(x,y):
             for i in range (len(x)):
                 print (x[i] + y[i])

In [39]: print ("Add two series\n")
         AddSeries (my_series_11, my_series1)
         Add two series
         10
         12
         14
         16
         18
         20
```

You can visualize data series using the different plotting systems that are covered in Chapter 7. However, Figure 6-1 demonstrates how to get an at-a-glance idea of your series data and graphically explore it via visual plotting diagrams. See Listing 6-9.

Listing 6-9. Visualizing Data Series

```
In [49]: import matplotlib.pyplot as plt
         plt.plot(my_series2)
         plt.ylabel('index')
         plt.show()
```

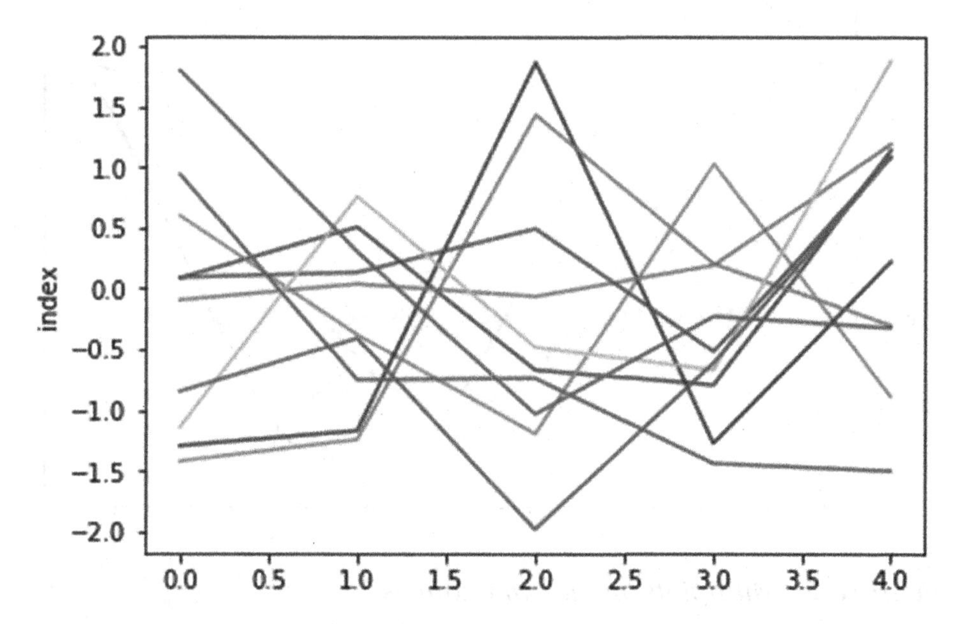

Figure 6-1. *Line visualization*

```
In [54]: from numpy import *
         import math
         import matplotlib.pyplot as plt
         t = linspace(0, 2*math.pi, 400)
```

```
        a = sin(t)
        b = cos(t)
        c = a + b

In [50]: plt.plot(t, a, 'r') # plotting t, a separately
         plt.plot(t, b, 'b') # plotting t, b separately
         plt.plot(t, c, 'g') # plotting t, c separately
         plt.show()
```

We can add multiple plots to the same canvas as shown in Figure 6-2.

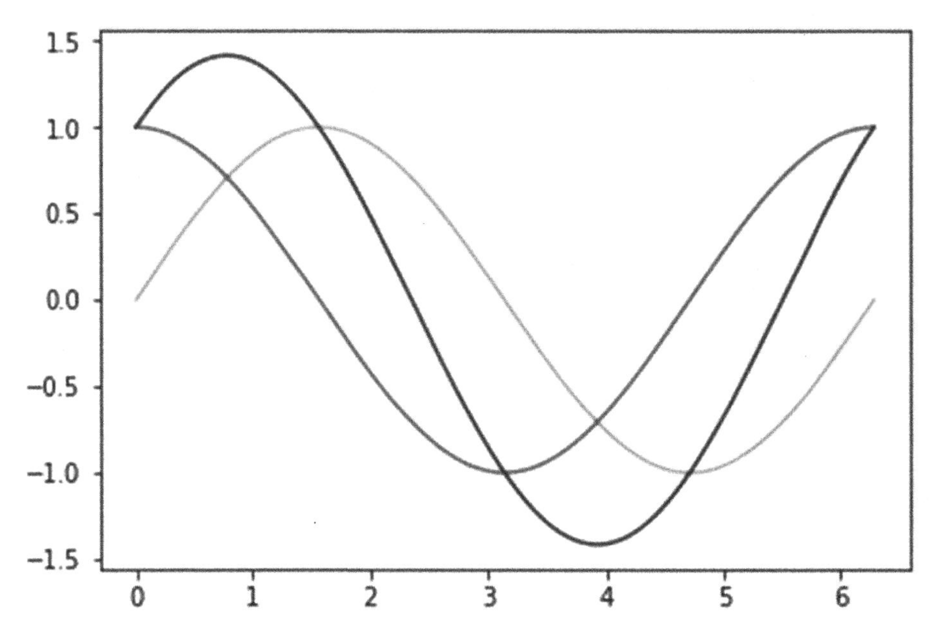

Figure 6-2. *Multiplots on the same canvas*

Data Frame Data Structures

As mentioned earlier, a data frame is a two-dimensional data structure with heterogeneous data types, i.e., tabular data.

Creating a Data Frame

Pandas can create a data frame using the constructor `pandas.DataFrame(data, index, columns, dtype, copy)`. A data frame can be created from lists, series, dictionaries, Numpy arrays, or other data frames. A Pandas data frame not only helps to store tabular data but also performs arithmetic operations on rows and columns of the data frame. Listing 6-10 creates a data frame from a single list and a list of lists.

Listing 6-10. Creating a Data Frame from a List

```
In [19]: import pandas as pd
         data = [10,20,30,40,50]
         DF1 = pd.DataFrame(data)
         print (DF1)
         0    10
         1    20
         2    30
         3    40
         4    50

In [22]: import pandas as pd
         data = [['Ossama',25],['Ali',43],['Ziad',32]]
         DF1 = pd.DataFrame(data,columns=['Name','Age'])
         print (DF1)
              Name       Age
         0    Ossama     25
         1    Ali        43
         2    Ziad       32

In [21]: import pandas as pd
         data = [['Ossama',25],['Ali',43],['Ziad',32]]
         DF1 = pd.DataFrame(data,columns=['Name','Age'],
         dtype=float) print (DF1)
```

```
    Name     Age
0   Ossama   25.0
1   Ali      43.0
2   Ziad     32.0
```

You can create a data frame from dictionaries or arrays, as shown in Listing 6-11. Also, you can set the data frame indices. However, if you don't set the indices, then the data frame starts with 0 and goes up to n-1, where n is the length of the list. Column names are taken by default from the dictionary keys. However, it's possible to set labels for columns as well. The first data frame's df1 columns are labeled with the dictionary key names; that's why you don't see NaN cases except for the missing value of the project in dictionary 1. While in the second data frame, named df2, you change the column name from Test1 to Test_1, and you get NaNs for all the records. This is because of the absence of Test_1 in the dictionary key of data.

Listing 6-11. Creating a DataFrame from a Dictionary

```
In [13]: import pandas as pd
         data = [{'Test1': 10, 'Test2': 20},{'Test1': 30,
                 'Test2': 20, 'Project': 20}]
         # With three column indices, values same as dictionary
         keys
         df1 = pd.DataFrame(data, index=['First', 'Second'],
         columns=['Test2', 'Project' , 'Test1'])

         #With two column indices with one index with another
         name
         df2 = pd.DataFrame(data, index=['First', 'Second'],
         columns=['Project', 'Test_1','Test2 ')]
         print (df1)
         print ("\n")
         print (df2)
```

```
        Test2    Project    Test1
First   20       NaN        10
Second  20       20.0       30

        Project  Test_1     Test2
First   NaN      NaN        20
Second  20.0     NaN        20
```

Pandas allows you to create a data frame from a dictionary of series where you get the union of all series indices passed. As shown in Listing 6-12 with the student `Salwa`, no `Test1` value is given. That's why NaN is set automatically.

Listing 6-12. Creating a Data Frame from a Series

```
In [16]: import pandas as pd
         data = {'Test1' : pd.Series([70, 55, 89],
                 index=['Ahmed', 'Omar', 'Ali']),
                 'Test2' : pd.Series([56, 82, 77, 65],
                 index=['Ahmed', 'Omar', 'Ali', 'Salwa'])}

         df1 = pd.DataFrame(data)
         print (df1)

                Test1    Test2
         Ahmed  70.0     56
         Ali    89.0     77
         Omar   55.0     82
         Salwa  NaN      65
```

Updating and Accessing a Data Frame's Column Selection

You can select a specific column using the column labels. For example, `df1['Test2']` is used to select only the column labeled `Test2` in the data frame, while `df1[:]` is used to display all the columns and all the rows, as shown in Listing 6-13.

Listing 6-13. Data Frame Column Selection

```
In [51]: import pandas as pd
         data = {'Test1' : pd.Series([70, 55, 89],
                 index=['Ahmed', 'Omar', 'Ali']),
                 'Test2' : pd.Series([56, 82, 77, 65],
                 index=['Ahmed', 'Omar', 'Ali', 'Salwa'])}

         df1 = pd.DataFrame(data)
         print (df1['Test2']) # Column selection
         print("\n")
         print (df1[:]) # Column selection

         Ahmed    56
         Ali      77
         Omar     82
         Salwa    65
         Name: Test2, dtype: int64

                  Test1    Test2
         Ahmed    70.0     56
         Ali      89.0     77
         Omar     55.0     82
         Salwa    NaN      65
```

You can select columns by using the column labels or the column index. `df1.iloc[:, [1,0]]` is used to display all rows for columns 1 and 0 starting with column 1, which refers to the column named `Test2`. In addition, `df1[0:4:1]` is used to display all the rows starting from row 0 up to row 3 incremented by 1, which gives all rows from 0 up to 3. See Listing 6-14.

Listing 6-14. Data Frame Column and Row Selection

```
In [46]: df1.iloc[:, [1,0 ]]

Out[46]:                  Test2   Test1
         Ahmed            56      70.0
         Ali              77      89.0
         Omar             82      55.0
         Salwa            65      NaN

In [39]: df1[0:4:1]

Out[39]:                  Test1   Test2
         Ahmed            70.0    56
         Ali              89.0    77
         Omar             55.0    82
         Salwa            NaN     65
```

Column Addition

You can simply add a new column and add its values directly using a series. In addition, you can create a new column by processing the other columns, as shown in Listing 6-15.

Listing 6-15. Adding a New Column to a Data Frame

```
In [66]: # add a new Column
         import pandas as pd
         data = {'Test1' : pd.Series([70, 55, 89],
                 index=['Ahmed', 'Omar', 'Ali']),
                 'Test2' : pd.Series([56, 82, 77, 65],
                 index=['Ahmed', 'Omar', 'Ali', 'Salwa'])}
              df1 = pd.DataFrame(data)
              print (df1)
              df1['Project'] = pd.Series([90,83,67, 87],
              index=['Ali','Omar','Salwa', 'Ahmed'])
              print ("\n")
         df1['Average'] = round((df1['Test1']+df1['Test2']+
         df1['Project'])/3, 2)
         print (df1)

                  Test1   Test2
         Ahmed    70.0    56
         Ali      89.0    77
         Omar     55.0    82
         Salwa    NaN     65

                  Test1   Test2   Project   Average
         Ahmed    70.0    56      87        71.00
         Ali      89.0    77      90        85.33
         Omar     55.0    82      83        73.33
         Salwa    NaN     65      67        NaN
```

Column Deletion

You can delete any column using the `del` method. For example, `del df2['Test2']` deletes the `Test2` column from the data set. In addition, you can use the pop method to delete a column. For example, df2.pop('Project') is used to delete the column Project. However, you should be careful when you use the del or pop method since a reference might exist. In this case, it deletes not only from the executed data frame but also from the referenced data frame. Listing 6-16 creates the data frame df1 and copies df1 to df2.

Listing 6-16. Creating and Copying a Data Frame

```
In [70]: import pandas as pd
         data = {'Test1' : pd.Series([70, 55, 89],
                 index=['Ahmed', 'Omar', 'Ali']),
                 'Test2' : pd.Series([56, 82, 77, 65],
                 index=['Ahmed', 'Omar', 'Ali', 'Salwa'])}
         print (df1)
         df2 = df1
         print ("\n")
         print (df2)
                  Test1    Test2    Project    Average
         Ahmed    70.0     56       87         71.00
         Ali      89.0     77       90         85.33
         Omar     55.0     82       83         73.33
         Salwa    NaN      65       67         NaN
                           Test1    Test2    Project    Average
Ahmed                      70.0     56       87         71.00
Ali                        89.0     77       90         85.33
Omar                       55.0     82       83         73.33
Salwa                      NaN      65       6          7  NaN
```

In the previous Python script, you saw how to create df2 and assign it df1. In Listing 6-17, you are deleting the Test2 and Project variables using the del and pop methods sequentially. As shown, both variables are deleted from both data frames df1 and df2 because of the reference existing between these two data frames as a result of using the assign (=) operator.

Listing 6-17. Deleting Columns from a Data Frame

```
In [71]: # Delete a column in data frame using del function
         print ("Deleting the first column using DEL function:")
         del df2['Test2']
         print (df2)
         # Delete a column in data frame using pop function
         print ("\nDeleting another column using POP function:")
         df2.pop('Project')
         print (df2)

         Deleting the first column using DEL function:

                     Test1   Project    Average
         Ahmed       70.0    87         71.00
         Ali         89.0    90         85.33
         Omar        55.0    83         73.33
         Salwa       NaN     67         NaN

         Deleting another column using POP function:
                     Test1    Average
         Ahmed       70.0     71.00
         Ali         89.0     85.33
         Omar        55.0     73.33
         Salwa       NaN      NaN

In [72]: print (df1)
                     Test1    Average
         Ahmed       70.0     71.00
         Ali         89.0     85.33
         Omar        55.0     73.33
         Salwa       NaN      NaN
```

```
In [73]: print (df2)
                    Test1     Average
         Ahmed      70.0      71.00
         Ali        89.0      85.33
         Omar       55.0      73.33
         Salwa      NaN       NaN
```

To solve this problem, you can use the `df. copy()` method instead of the assign operator (=). Listing 6-18 shows that you deleted the variables `Test2` and `Project` using the `del()` and `pop()` methods sequentially, but only `df2` has been affected, while `df1` remains unchanged.

Listing 6-18. Using the Copy Method to Delete Columns from a Data Frame

```
In [83]: # add a new Column
         import pandas as pd
         data = {'Test1' : pd.Series([70, 55, 89],
                 index=['Ahmed', 'Omar', 'Ali']),
                 'Test2' : pd.Series([56, 82, 77, 65],
                 index=['Ahmed', 'Omar', 'Ali', 'Salwa'])}
         df1 = pd.DataFrame(data)
         df1['Project'] = pd.Series([90,83,67, 87],
         index=['Ali','Omar','Salwa', 'Ahmed'])
         print ("\n")
         df1['Average'] = round((df1['Test1']+df1['Test2']+df1
         ['Project'])/3, 2)
         print (df1)
         print ("\n")
         df2= df1.copy() # copy df1 into df2 using copy() method
         print (df2)
         #delete columns using del and pop methods
         del df2['Test2']
```

```
df2.pop('Project')
print ("\n")
print (df1)
print ("\n")
print (df2)
```

```
       Test1  Test2  Project  Average
Ahmed   70.0     56       87    71.00
Ali     89.0     77       90    85.33
Omar    55.0     82       83    73.33
Salwa    NaN     65       67      NaN


       Test1  Test2  Project  Average
Ahmed   70.0     56       87    71.00
Ali     89.0     77       90    85.33
Omar    55.0     82       83    73.33
Salwa    NaN     65       67      NaN


       Test1  Test2  Project  Average
Ahmed   70.0     56       87    71.00
Ali     89.0     77       90    85.33
Omar    55.0     82       83    73.33
Salwa    NaN     65       67      NaN


       Test1  Average
Ahmed   70.0    71.00
Ali     89.0    85.33
Omar    55.0    73.33
Salwa    NaN      NaN
```

Row Selection

In Listing 6-19, you are selecting the second row for student `Omar`. Also, you use the slicing methods to retrieve rows 2 and 3.

Listing 6-19. Retrieving Specific Rows

```
In [106]: # add a new Column
          import pandas as pd
          data = {'Test1' : pd.Series([70, 55, 89],
                  index=['Ahmed', 'Omar', 'Ali']),
                  'Test2' : pd.Series([56, 82, 77, 65],
                  index=['Ahmed', 'Omar', 'Ali', 'Salwa'])}
          df1 = pd.DataFrame(data)
          df1['Project'] = pd.Series([90,83,67, 87],index=
          ['Ali','Omar','Salwa', 'Ahmed'])
         print ("\n")
         df1['Average'] = round((df1['Test1']+df1['Test2']+df1
         ['Project'])/3, 2)
         print (df1)
         print ("\nselect iloc function to retrieve row number 2")
         print (df1.iloc[2])
         print ("\nslice rows")
         print (df1[2:4] )
```

```
       Test1  Test2  Project  Average
Ahmed   70.0     56       87    71.00
Ali     89.0     77       90    85.33
Omar    55.0     82       83    73.33
Salwa    NaN     65       67      NaN

select  iloc function to retrieve  row number 2
Test1       55.00
Test2       82.00
Project     83.00
Average     73.33
Name: Omar, dtype: float64

slice rows
       Test1  Test2  Project  Average
Omar    55.0     82       83    73.33
Salwa    NaN     65       67      NaN
```

Row Addition

Listing 6-20 demonstrates how to add rows to an existing data frame.

Listing 6-20. Adding New Rows to the Data Frame

```
In [134 ]: import pandas as pd
           data = {'Test1' : pd.Series([70, 55, 89],
                   index=['Ahmed', 'Omar', 'Ali']),
                   'Test2' : pd.Series([56, 82, 77, 65],
                   index=['Ahmed', 'Omar', 'Ali', 'Salwa']),
                'Project' : pd.Series([87, 83, 90, 67],
                index=['Ahmed', 'Omar', 'Ali', 'Salwa']),
                'Average' : pd.Series([71, 73.33, 85.33, 66],
                index=['Ahmed', 'Omar', 'Ali', 'Salw
           data = pd.DataFrame(data)
           print (data)
           print("\n")
           df2 = pd.DataFrame([[80, 70, 90, 80]], columns
           = ['Test1','Test2','Project','Average'],
           index=['Khalid'])
           datadata.append(df2)
           print (data)
```

```
        Average  Project  Test1  Test2
Ahmed     71.00       87   70.0     56
Ali       85.33       90   89.0     77
Omar      73.33       83   55.0     82
Salwa     66.00       67    NaN     65

         Average  Project  Test1  Test2
Ahmed      71.00       87   70.0     56
Ali        85.33       90   89.0     77
Omar       73.33       83   55.0     82
Salwa      66.00       67    NaN     65
Khalid     80.00       90   80.0     70
```

Row Deletion

Pandas provides the `df.drop()` method to delete rows using the label index, as shown in Listing 6-21.

Listing 6-21. Deleting Rows from a Data Frame

```
In [138]: print (data)
       print ('\n')
       data = data.drop('Omar')
       print (data)
```

```
        Average  Project  Test1  Test2
Ahmed     71.00       87   70.0     56
Ali       85.33       90   89.0     77
Omar      73.33       83   55.0     82
Salwa     66.00       67    NaN     65
Khalid    80.00       90   80.0     70


        Average  Project  Test1  Test2
Ahmed     71.00       87   70.0     56
Ali       85.33       90   89.0     77
Salwa     66.00       67    NaN     65
Khalid    80.00       90   80.0     70
```

Exploring and Analyzing a Data Frame

Pandas provides various methods for analyzing data in a data frame. The `.describe()` method is used to generate descriptive statistics that summarize the central tendency, dispersion, and shape of a data set's distribution, excluding NaN values.

```
DataFrame.describe(percentiles=None,include=None, exclude=None)
[source]
```

`DataFrame.describe()` analyzes both numeric and object series, as well as data frame column sets of mixed data types. The output will vary depending on what is provided. Listing 6-22 analyzes the `Age`, `Salary`,

Height, and Weight attributes in a data frame. It also shows the mean, max, min, standard deviation, and quantiles of all attributes. However, Salwa's Age is missing; you get the full description of Age attributes excluding Salwa's data.

Listing 6-22. Creating a Data Frame with Five Attributes

```
In [61]: print (df1)
data = {'Age' : pd.Series([30, 25, 44, ],
index=['Ahmed', 'Omar', 'Ali']),
'Salary' : pd.Series([25000, 17000, 30000, 12000],
index=['Ahmed', 'Omar', 'Ali',
'Height' : pd.Series([160, 154, 175, 165],
index=['Ahmed', 'Omar', 'Ali', 'Salwa'
'Weight' : pd.Series([85, 70, 92, 65], index=['Ahmed', 'Omar',
'Ali', 'Salwa']),
'Gender' : pd.Series(['Male', 'Male', 'Male', 'Female'],
index=['Ahmed', 'Omar',

data = pd.DataFrame(data)
print (data)
print("\n")
df2 = pd.DataFrame([[42, 31000, 170, 80, 'Female']], columns
=['Age','Salary','Height'
                          , index=['Mona'])

data = data.append(df2)
print (data)
```

```
       Age  Gender  Height  Salary  Weight
Ahmed  30.0   Male     160   25000      85
Ali    44.0   Male     175   30000      92
Omar   25.0   Male     154   17000      70
Salwa   NaN Female     165   12000      65


       Age  Gender  Height  Salary  Weight
Ahmed  30.0   Male     160   25000      85
Ali    44.0   Male     175   30000      92
Omar   25.0   Male     154   17000      70
Salwa   NaN Female     165   12000      65
Mona   42.0 Female     170   31000      80
```

Applying the `data.describe()` method, you get the full description of all attributes except the `Gender` attribute because of its string data type. You can enforce implementation of all attributes by using the `include='all'` method attribute. Also, you can apply the analysis to a specific pattern, for example, to the `Salary` pattern only, which finds the mean, min, max, std, and quantiles of all employees' salaries. See Listing 6-23.

Listing 6-23. Analyzing a Data Frame

```
In [63]: data.describe()
```

Out[63]:

	Age	Height	Salary	Weight
count	4.000000	5.000000	5.000000	5.000000
mean	35.250000	144.800000	23000.000000	78.400000
std	9.215024	42.517055	8276.472679	10.968136
min	25.000000	70.000000	12000.000000	65.000000
25%	28.750000	154.000000	17000.000000	70.000000
50%	36.000000	160.000000	25000.000000	80.000000
75%	42.500000	165.000000	30000.000000	85.000000
max	44.000000	175.000000	31000.000000	92.000000

```
In [64]: data.describe(include='all')
```

Out[64]:

	Age	Gender	Height	Salary	Weight
count	4.000000	5	5.000000	5.000000	5.000000
unique	NaN	2	NaN	NaN	NaN
top	NaN	Male	NaN	NaN	NaN
freq	NaN	3	NaN	NaN	NaN
mean	35.250000	NaN	144.800000	23000.000000	78.400000
std	9.215024	NaN	42.517055	8276.472679	10.968136
min	25.000000	NaN	70.000000	12000.000000	65.000000
25%	28.750000	NaN	154.000000	17000.000000	70.000000
50%	36.000000	NaN	160.000000	25000.000000	80.000000
75%	42.500000	NaN	165.000000	30000.000000	85.000000
max	44.000000	NaN	175.000000	31000.000000	92.000000

```
In [66]: data.Salary.describe()

Out[66]: count        5.000000
         mean     23000.000000
         std       8276.472679
         min      12000.000000
         25%      17000.000000
         50%      25000.000000
         75%      30000.000000
         max      31000.000000
         Name: Salary, dtype: float64
```

Listing 6-24 includes only the numeric columns in a data frame's description.

Listing 6-24. Analyzing Only Numerical Patterns

```
In [67]: data.describe(include=[np.number])
```

Out[67]:

	Age	Height	Salary	Weight
count	4.000000	5.000000	5.000000	5.000000
mean	35.250000	144.800000	23000.000000	78.400000
std	9.215024	42.517055	8276.472679	10.968136
min	25.000000	70.000000	12000.000000	65.000000
25%	28.750000	154.000000	17000.000000	70.000000
50%	36.000000	160.000000	25000.000000	80.000000
75%	42.500000	165.000000	30000.000000	85.000000
max	44.000000	175.000000	31000.000000	92.000000

Listing 6-25 includes only string columns in a data frame's description.

Listing 6-25. Analyzing String Patterns Only (Gender)

```
In [68]: data.describe(include=[np.object])
```

Out[68]:

	Gender
count	5
unique	2
top	Male
freq	3

```
In [70]: data.describe(exclude=[np.number])
```

Out[70]:

	Gender
count	5
unique	2
top	Male
freq	3

You can measure overweight employee by calculating the optimal weight and comparing this with their recorded weight, as shown in Listing 6-26.

Listing 6-26. Checking the Weight Optimality

```
In [71]: data
```

Out[71]:

	Age	Gender	Height	Salary	Weight
Ahmed	30.0	Male	160	25000	85
Ali	44.0	Male	175	30000	92
Omar	25.0	Male	154	17000	70
Salwa	NaN	Female	165	12000	65
Mona	42.0	Female	70	31000	80

```
In [75]: OptimalWeight = data['Height']- 100
         OptimalWeight

Out[75]: Ahmed     60
         Ali       75
         Omar      54
         Salwa     65
         Mona      70
         Name: Height, dtype: int64

In [93]:unOptimalCases = data['Weight'] <= OptimalWeight
unOptimalCases

Out[93]: Ahmed     False
         Ali       False
         Omar      False
         Salwa      True
         Mona      False
         dtype: bool
```

Panel Data Structures

As mentioned earlier, a *panel* is a three-dimensional data structure like a three-dimensional array.

Creating a Panel

Pandas creates a panel using the constructor `pandas.Panel(data, items, major_axis, minor_axis, dtype, copy)`. The panel can be created from a dictionary of data frames and narrays. The data can take various forms, such as ndarray, series, map, lists, dictionaries, constants, and also another data frames.

The following Python script creates an empty panel:

```
#creating an empty panel
import pandas as pd
p = pd.Panel ()
```

Listing 6-27 creates a panel with three dimensions.

Listing 6-27. Creating a Panel with Three Dimensions

```
In [143]: # creating an empty panel
          import pandas as pd
          import numpy as np

          data = np.random.rand(2,4,5)
          Paneldf = pd.Panel(data)
          print (Paneldf)

<class 'pandas.core.panel.Panel'>
Dimensions: 2 (items) x 4 (major_axis) x 5 (minor_axis)
Items axis: 0 to 1
Major_axis axis: 0 to 3
Minor_axis axis: 0 to 4
```

Accessing Data from a Panel with a Position

Listing 6-28 creates a panel and fills it with random data, where the first item in the panel is a 4x3 array and the second item is a 4x2 array of random values. For the `Item2` column, two values are NaN since its dimension is 4x2. You can also access data from a panel using item labels, as shown in Listing 6-28.

Listing 6-28. Selecting and Displaying Panel Items

```
In [147]: # creating an empty panel

import pandas as pd
import numpy as np
data = {'Item1' : pd.DataFrame(np.random.randn(4, 3)),
        'Item2' : pd.DataFrame(np.random.randn(4, 2))}
Paneldf = pd.Panel(data)
print (Paneldf['Item1'])
print ("\n")
print (Paneldf['Item2'])
```

```
          0         1         2
0 -1.069595  0.835842  0.950269
1  1.063784  0.520086  1.342309
2 -2.236069  0.229717  0.752612
3  1.014550  0.903234  2.011993


          0         1   2
0 -1.126333  1.528085 NaN
1 -1.255712  0.076873 NaN
2  1.593704 -0.648342 NaN
3  0.287446  1.591275 NaN
```

Python displays the panel items in a data frame with two dimensions, as shown previously. Data can be accessed using the method `panel.major_axis(index)` and also using the method `panel.minor_axis(index)`. See Listing 6-29.

Listing 6-29. Selecting and Displaying a Panel with Major and Minor Dimensions

```
In [149]: print (Paneldf.major_xs(1))
```

```
      Item1      Item2
0  1.063784 -1.255712
1  0.520086  0.076873
2  1.342309        NaN
```

```
In [150]: print (Paneldf.minor_xs(1))
```

```
      Item1     Item2
0  0.835842  1.528085
1  0.520086  0.076873
2  0.229717 -0.648342
3  0.903234  1.591275
```

Exploring and Analyzing a Panel

Once you have a panel, you can make statistical analysis on the maintained data. In Listing 6-30, you can see two groups of employees, each of which has five attributes maintained in a panel called P. You implement the .describe() method for Group1, as well as for the Salary attribute in this group.

Listing 6-30. Panel Analysis

```
In [104]: import pandas as pd
data1 = {'Age' : pd.Series([30, 25, 44, ], index=['Ahmed',
'Omar', 'Ali']),
'Salary' : pd.Series([25000, 17000, 30000, 12000],
index=['Ahmed', 'Omar', 'Ali', 'Salwa']),
'Height' : pd.Series([160, 154, 175, 165], index=['Ahmed',
'Omar', 'Ali', 'Salwa']),
```

```
'Weight' : pd.Series([85, 70, 92, 65], index=['Ahmed', 'Omar',
'Ali', 'Salwa']),
'Gender' : pd.Series(['Male', 'Male', 'Male', 'Female'],
index=['Ahmed', 'Omar', 'Ali', 'Salwa'])}

data2 = {'Age' : pd.Series([24, 19, 33,25  ], index=['Ziad',
'Majid', 'Ayman', 'Ahlam']),
'Salary' : pd.Series([17000, 7000, 22000, 21000],
index=['Ziad', 'Majid', 'Ayman', 'Ahlam']),
'Height' : pd.Series([170, 175, 162, 177], index=['Ziad',
'Majid', 'Ayman', 'Ahlam']),
'Weight' : pd.Series([77, 84, 74, 90], index=['Ziad', 'Majid',
'Ayman', 'Ahlam']),
'Gender' : pd.Series(['Male', 'Male', 'Male', 'Female'],
index=['Ziad', 'Majid', 'Ayman', 'Ahlam'])}

data = {'Group1': data1, 'Group2': data2}
p = pd.Panel(data)

In [106]: p['Group1'].describe()
```

Out[106]:

	Age	Gender	Height	Salary	Weight
count	3.0	4	4.0	4.0	4.0
unique	3.0	2	4.0	4.0	4.0
top	30.0	Male	175.0	30000.0	70.0
freq	1.0	3	1.0	1.0	1.0

```
In [107]: p['Group1']['Salary'].describe()

Out[107]: count          4.0
          unique         4.0
          top        30000.0
          freq           1.0
          Name: Salary, dtype: float64
```

Data Analysis

As indicated earlier, Pandas provides numerous methods for data analysis. The objective in this section is to get familiar with the data and summarize its main characteristics. Also, you can define your own methods for specific statistical analyses.

Statistical Analysis

Most of the following statistical methods were covered earlier with practical examples of the three main data collections: series, data frames, and panels.

- `df.describe()`: Summary statistics for numerical columns
- `df.mean()`: Returns the mean of all columns
- `df.corr()`: Returns the correlation between columns in a data frame
- `df.count()`: Returns the number of non-null values in each data frame column
- `df.max()`: Returns the highest value in each column
- `df.min()`: Returns the lowest value in each column
- `df.median()`: Returns the median of each column
- `df.std()`: Returns the standard deviation of each column

Listing 6-31 creates a data frame with six columns and ten rows.

Listing 6-31. Creating a Data Frame

```
In [11]: import pandas as pd
import numpy as np
```

```
Number = [1,2,3,4,5,6,7,8,9,10]
Names = ['Ali Ahmed','Mohamed Ziad','Majid Salim','Salwa
Ahmed', 'Ahlam Mohamed', 'Omar Ali', 'Amna Mohammed','Khalid
Yousif', 'Safa Humaid', 'Amjad Tayel']
City = ['Fujairah','Dubai','Sharjah','AbuDhabi','Fujairah','Dub
ai', 'Sharja ', 'AbuDhabi','Sharjah','Fujairah']
columns = ['Number', 'Name', 'City' ]
dataset= pd.DataFrame({'Number': Number , 'Name': Names,
'City': City}, columns = columns )
Gender= pd.DataFrame({'Gender':['Male','Male','Male','Female',
'Female', 'Male', 'Female', 'Male','Female', 'Male']})
Height = pd.DataFrame(np.random.randint(120,175, size=(12, 1)))
Weight = pd.DataFrame(np.random.randint(50,110, size=(12, 1)))
dataset['Gender']= Gender
dataset['Height']= Height
dataset['Weight']= Weight
dataset.set_index('Number')
```

Out[166]:

Number	Name	City	Gender	Height	Weight
1	Ali Ahmed	Fujairah	Male	131	71
2	Mohamed Ziad	Dubai	Male	153	74
3	Majid Salim	Sharjah	Male	145	104
4	Salwa Ahmed	AbuDhabi	Female	173	86
5	Ahlam Mohamed	Fujairah	Female	158	82
6	Omar Ali	Dubai	Male	134	89
7	Amna Mohammed	Sharjah	Female	136	93
8	Khalid Yousif	AbuDhabi	Male	128	98
9	Safa Humaid	Sharjah	Female	162	81
10	Amjad Tayel	Fujairah	Male	160	77

The Python script and examples in Listing 6-32 show the summary of height and weight variables, the mean values of height and weight, the correlation between the numerical variables, and the count of all records in the data set. The correlation coefficient is a measure that determines the degree to which two variables' movements are associated. The most common correlation coefficient, generated by the Pearson correlation, may be used to measure the linear relationship between two variables. However, in a nonlinear relationship, this correlation coefficient may not always be a suitable measure of dependence. The range of values for the correlation coefficient is -1.0 to 1.0. In other words, the values cannot exceed 1.0 or be less than -1.0, whereby a correlation of -1.0 indicates a perfect negative correlation, and a correlation of 1.0 indicates a perfect positive correlation. The correlation coefficient is denoted as `r`. If its value greater than zero, it's a positive relationship; while if the value is less than zero, it's a negative relationship. A value of zero indicates that there is no relationship between the two variables.

As shown, there is a weak negative correlation (-0.301503) between the height and width of all members in the data set. Also, the initial stats show that the height has the highest deviation; in addition, the 75th quantile of the height is equal to 159.

Listing 6-32. Summary and Statistics of Variables

```
In [186]: # Summary statistics for numerical columns
print ( dataset.describe())
```

```
          Number     Height      Weight
count  10.00000   10.00000   10.000000
mean    5.50000  148.00000   85.500000
std     3.02765   15.37675   10.617072
min     1.00000  128.00000   71.000000
25%     3.25000  134.50000   78.000000
50%     5.50000  149.00000   84.000000
75%     7.75000  159.50000   92.000000
max    10.00000  173.00000  104.000000
```

```
In [187]: print (dataset.mean()) # Returns the mean of all
columns

Number      5.5
Height    148.0
Weight     85.5
dtype: float64


In [188]: # Returns the correlation between columns in a
DataFrame
print (dataset.corr())

          Number    Height    Weight
Number  1.000000  0.124105  0.174557
Height  0.124105  1.000000 -0.301503
Weight  0.174557 -0.301503  1.000000


In [189]: # Returns the number of non-null values in each
DataFrame      column
print (dataset.count())

Number    10
Name      10
City      10
Gender    10
Height    10
Weight    10
dtype: int64
```

```
In [190]: # Returns the highest value in each column
print (dataset.max())
```

```
Number              10
Name       Salwa Ahmed
City           Sharjah
Gender            Male
Height             173
Weight             104
dtype: object
```

```
In [191]: # Returns the lowest value in each column
print (dataset.min())
```

```
Number                1
Name      Ahlam Mohamed
City           AbuDhabi
Gender           Female
Height              128
Weight               71
dtype: object
```

```
In [192]: # Returns the median of each column
print (dataset.median())
```

```
Number      5.5
Height    149.0
Weight     84.0
dtype: float64
```

```
In [193]: # Returns the standard deviation of each column
print (dataset.std())
```

```
Number     3.027650
Height    15.376750
Weight    10.617072
dtype: float64
```

Data Grouping

You can split data into groups to perform more specific analysis over the data set. Once you perform data grouping, you can compute summary statistics (aggregation), perform specific group operations (transformation), and discard data with some conditions (filtration). In Listing 6-33, you group data using `City` and find the count of genders per city. In addition, you group the data set by city and display the results, where for example rows 1 and 5 are people from Dubai. You can use multiple grouping attributes. You can group the data set using `City` and `Gender`. The retrieved data shows that, for instance, Fujairah has females (row 4) and males (rows 0 and 9).

Listing 6-33. Data Grouping

```
In [3]: dataset.groupby('City')['Gender'].count()
```

The following output shows that we have 2 students from Abu dhabi, 2 from Dubai, 3 from Fujairah and 3 from Sharjah groupped by gender.

```
Out[3]: City
        AbuDhabi    2
        Dubai       2
        Fujairah    3
        Sharjah     3
        Name: Gender, dtype: int64
```

```
In [4]: print (dataset.groupby('City').groups)
```

```
{'AbuDhabi': Int64Index([3, 7], dtype='int64'), 'Dubai': Int64Index([1, 5], dtype='int64'), 'Fujairah': Int64I
ndex([0, 4, 9], dtype='int64'), 'Sharjah': Int64Index([2, 6, 8], dtype='int64')}
```

```
In [5]: print (dataset.groupby(['City','Gender']).groups)
```

```
{('AbuDhabi', 'Female'): Int64Index([3], dtype='int64'), ('AbuDhabi', 'Male'): Int64Index([7], dtype='int64'),
('Dubai', 'Male'): Int64Index([1, 5], dtype='int64'), ('Fujairah', 'Female'): Int64Index([4], dtype='int64'),
('Fujairah', 'Male'): Int64Index([0, 9], dtype='int64'), ('Sharjah', 'Female'): Int64Index([6, 8], dtype='int6
4'), ('Sharjah', 'Male'): Int64Index([2], dtype='int64')}
```

Iterating Through Groups

You can iterate through a specific group, as shown in Listing 6-34. When you iterate through the gender, it should be clear that by default the groupby object has the same name as the group name.

Listing 6-34. Iterating Through Grouped Data

```
In [7]: grouped = dataset.groupby('Gender')
        for name,group in grouped:
            print (name)
            print (group)
            print ("\n")
```

```
Female
   Number           Name      City  Gender  Height  Weight
3       4    Salwa Ahmed  AbuDhabi  Female     125      57
4       5  Ahlam Mohamed  Fujairah  Female     170      99
6       7  Amna Mohammed   Sharjah  Female     160      97
8       9    Safa Humaid   Sharjah  Female     138      70


Male
   Number           Name      City Gender  Height  Weight
0       1      Ali Ahmed  Fujairah   Male     130      72
1       2   Mohamed Ziad     Dubai   Male     129      61
2       3    Majid Salim   Sharjah   Male     153      51
5       6       Omar Ali     Dubai   Male     135      97
7       8  Khalid Yousif  AbuDhabi   Male     170      55
9      10    Amjad Tayel  Fujairah   Male     163      88
```

You can also select a specific group using the get_group() method, as shown in Listing 6-35 where you group data by gender and then select only females.

Listing 6-35. Selecting a Single Group

```
In [9]: grouped = dataset.groupby('Gender')
        print (grouped.get_group('Female'))
```

```
   Number           Name      City  Gender  Height  Weight
3       4    Salwa Ahmed  AbuDhabi  Female     125      57
4       5  Ahlam Mohamed  Fujairah  Female     170      99
6       7  Amna Mohammed   Sharjah  Female     160      97
8       9    Safa Humaid   Sharjah  Female     138      70
```

Aggregations

Aggregation functions return a single aggregated value for each group. Once the groupby object is created, you can implement various functions on the grouped data. In Listing 6-36, you calculate the mean and size of height and weight for both males and females. In addition, you calculate the summation and standard deviations for both patterns of males and females.

Listing 6-36. Data Aggregation

```
In [18]: # Aggregation
         grouped = dataset.groupby('Gender')
         print (grouped['Height'].agg(np.mean))
         print ("\n")
         print (grouped['Weight'].agg(np.mean))
         print ("\n")
         print (grouped.agg(np.size))
         print ("\n")
         print (grouped['Height'].agg([np.sum, np.mean,
         np.std]))
```

```
Gender
Female     145.250000
Male       159.333333
Name: Height, dtype: float64

Gender
Female     88.750000
Male       83.666667
Name: Weight, dtype: float64

        Number  Name  City  Height  Weight
Gender
Female       4     4     4       4       4
Male         6     6     6       6       6

        sum        mean       std
Gender
Female  581  145.250000  7.274384
Male    956  159.333333  8.891944
```

Transformations

Transformation on a group or a column returns an object that is indexed the same size as the one being grouped. Thus, the transform should return a result that is the same size as that of a group chunk. See Listing 6-37.

Listing 6-37. Creating the Index

```
In [26]: dataset = dataset.set_index(['Number'])
         print (dataset)

                 Name      City  Gender  Height  Weight
Number
1          Ali Ahmed  Fujairah    Male     155      65
2       Mohamed Ziad     Dubai    Male     165      59
3        Majid Salim   Sharjah    Male     159      82
4        Salwa Ahmed  AbuDhabi  Female     138     106
5      Ahlam Mohamed  Fujairah  Female     152     100
6           Omar Ali     Dubai    Male     145     108
7      Amna Mohammed   Sharjah  Female     151      67
8      Khalid Yousif  AbuDhabi    Male     171      96
9        Safa Humaid   Sharjah  Female     140      82
10       Amjad Tayel  Fujairah    Male     161      92
```

In Listing 6-38, you group data by Gender, then implement the function lambda x: (x - x.mean()) / x.std()*10, and display results for both height and weight. The lambda operator or lambda function is a way to create a small anonymous function, i.e., a function without a name. This function is throwaway function; in other words, it is just needed where it has been created.

Listing 6-38. Transformation

```
In [28]: grouped = dataset.groupby('Gender')
         score = lambda x: (x - x.mean()) / x.std()*10
         print (grouped.transform(score))
```

```
          Height     Weight
Number
1       -4.873325  -9.911893
2        6.372810 -13.097858
3       -0.374871  -0.884990
4       -9.966479   9.730865
5        9.279136   6.346216
6      -16.119460  12.920860
7        7.904449 -12.269352
8       13.120491   6.548929
9       -7.217106  -3.807730
10       1.874356   4.424952
```

Filtration

Python provides direct filtering for data. In Listing 6-39, you applied filtering by city, and the return cities appear more than three times in the data set.

Listing 6-39. Filtration

```
In [30]: print (dataset.groupby('City').filter(lambda x: len(x)
>= 3))
```

```
                Name      City  Gender  Height  Weight
Number
1          Ali Ahmed  Fujairah    Male     155      65
3        Majid Salim   Sharjah    Male     159      82
5      Ahlam Mohamed  Fujairah  Female     152     100
7      Amna Mohammed   Sharjah  Female     151      67
9        Safa Humaid   Sharjah  Female     140      82
10       Amjad Tayel  Fujairah    Male     161      92
```

Summary

This chapter covered how to explore and analyze data in different collection structures. Here is a list of what you just studied in this chapter:

- How to implement Python techniques to explore and analyze a series of data, create a series, access data from series with the position, and apply statistical methods on a series.
- How to explore and analyze data in a data frame, create a data frame, and update and access data. This included column and row selection, addition, and deletion, as well as applying statistical methods on a data frame.
- How to apply statistical methods on a panel to explore and analyze its data.
- How to apply statistical analysis on the derived data from implementing Python data grouping, iterating through groups, aggregations, transformations, and filtration techniques.

The next chapter will cover how to visualize data using numerous plotting packages and much more.

Exercises and Answers

A. Create a data frame called `df` from the following tabular data dictionary that has these index labels: `['a', 'b', 'c', 'd', 'e', 'f', 'g', 'h', 'i', 'j']`.

	Animal	Age	Priority	Visits
a	cat	2.5	yes	1
b	cat	3.0	yes	3
c	snake	0.5	no	2
d	dog	NaN	yes	3
e	dog	5.0	no	2
f	cat	2.0	no	3
g	snake	4.5	no	1
h	cat	NaN	yes	1
i	dog	7.0	no	2
j	dog	3.0	no	1

Answer:

You should import both the Pandas and Numpy libraries.

```
import numpy as np
import pandas as pd
```

You must create a dictionary and list of labels and then call the data frame method and assign the labels list as an index, as shown in Listing 6-40.

Listing 6-40. Creating a Tabular Data Frame

```
In [5]: import numpy as np
        import pandas as pd
        import matplotlib as mpl
```

```
data = { 'Animal': ['cat', 'cat', 'snake', 'dog', 'dog',
                 'cat', 'snake', 'cat', 'dog', 'dog'],
'Age': [2.5, 3, 0.5, np.nan, 5, 2, 4.5, np.nan, 7, 3],
'Visits': [1, 3, 2, 3, 2, 3, 1, 1, 2, 1],
'Priority': ['yes', 'yes', 'no', 'yes', 'no', 'no', 'no',
'yes', 'no', 'no']}

labels = ['a', 'b', 'c', 'd', 'e', 'f', 'g', 'h', 'i', 'j']

#Create a DataFrame df from this dictionary data which has the
index labels.
df = pd.DataFrame( data, index = labels, columns=['Animal',
'Age', 'Priority', 'Visits'])
print (df)
```

```
  Animal  Age Priority  Visits
a    cat  2.5      yes       1
b    cat  3.0      yes       3
c  snake  0.5       no       2
d    dog  NaN      yes       3
e    dog  5.0       no       2
f    cat  2.0       no       3
g  snake  4.5       no       1
h    cat  NaN      yes       1
i    dog  7.0       no       2
j    dog  3.0       no       1
```

B. Display a summary of the data frame's basic information.

You can use `df.info()` and `df.describe()` to get a full description of your data set, as shown in Listing 6-41.

Listing 6-41. Data Frame Summary

```
In [6]: df.info()
```

```
<class 'pandas.core.frame.DataFrame'>
Index: 10 entries, a to j
Data columns (total 4 columns):
Animal      10 non-null object
Age         8 non-null float64
Priority    10 non-null object
Visits      10 non-null int64
dtypes: float64(1), int64(1), object(2)
memory usage: 400.0+ bytes
```

```
In [7]: df.describe()
```

	Age	Visits
count	8.000000	10.000000
mean	3.437500	1.900000
std	2.007797	0.875595
min	0.500000	1.000000
25%	2.375000	1.000000
50%	3.000000	2.000000
75%	4.625000	2.750000
max	7.000000	3.000000

C. Return the first three rows of the data frame `df`.

Listing 6-42 shows the use of `df.iloc[:3]` and `df.head(3)` to retrieve the first n rows of the data frame.

Listing 6-42. Selecting a Specific n Rows

```
In [12]: df.head(3)
```

Out[12]:

	Animal	Age	Priority	Visits
a	cat	2.5	yes	1
b	cat	3.0	yes	3
c	snake	0.5	no	2

```
In [13]: df.iloc[:3]
```

Out[13]:

	Animal	Age	Priority	Visits
a	cat	2.5	yes	1
b	cat	3.0	yes	3
c	snake	0.5	no	2

D. Select just the `animal` and `age` columns from the data frame `df`.

The Python data frame `loc()` method is used to retrieve the specific pattern `df.loc[ : , ['Animal', 'Age']]`. In addition, an array form retrieval can be used too with `df[['Animal', 'Age']]` . See Listing 6-43.

Listing 6-43. Slicing Data Frame

```
In [16]: df.loc[:,['Animal', 'Age']]
         # or
         df [['Animal', 'Age']]
```

Out[16]:

	Animal	Age
a	cat	2.5
b	cat	3.0
c	snake	0.5
d	dog	NaN
e	dog	5.0
f	cat	2.0
g	snake	4.5
h	cat	NaN
i	dog	7.0
j	dog	3.0

E. Count the visit priority per animal.

```
In [8]: df.groupby('Priority')['Animal'].count()
```

F. Find the mean of the animals' ages.

```
In [10]: df.groupby('Animal')['Age'].mean()
```

G. Display a summary of the data set. See Listing 6-44.

Listing 6-44. Data Set Summary

```
In [13]: df.groupby('Animal')['Age'].describe()
```

Out[13]:

	count	mean	std	min	25%	50%	75%	max
Animal								
cat	3.0	2.5	0.500000	2.0	2.25	2.5	2.75	3.0
dog	3.0	5.0	2.000000	3.0	4.00	5.0	6.00	7.0
snake	2.0	2.5	2.828427	0.5	1.50	2.5	3.50	4.5

CHAPTER 7

Data Visualization

Python provides numerous methods for data visualization. Various Python libraries can be used for data visualization, such as Pandas, Seaborn, Bokeh, Pygal, and Ploty. Python Pandas is the simplest method for basic plotting. Python Seaborn is great for creating visually appealing statistical charts that include color. Python Bokeh works great for more complicated visualizations, especially for web-based interactive presentations. Python Pygal works well for generating vector and interactive files. However, it does not have the flexibility that other methods do. Python Plotly is the most useful and easiest option for creating highly interactive web-based visualizations.

Bar charts are an essential visualization tool used to compare values in a variety of categories. A bar chart can be vertically or horizontally oriented by adjusting the x- and y-axes, depending on what kind of information or categories the chart needs to present. This chapter demonstrates the use and implementation of various visualization tools; the chapter will use the `salaries.csv` file shown in Figure 7-1 as the data set for plotting purposes.

O. Embarak, *Data Analysis and Visualization Using Python*,
https://doi.org/10.1007/978-1-4842-4109-7_7

rank	discipline	phd	service	sex	salary
AssocProf	B	11	11	Female	103613
Prof	A	12	6	Male	93000
Prof	A	23	20	Male	110515
Prof	A	40	31	Male	131205
Prof	B	20	18	Male	104800
Prof	A	20	20	Male	122400
AssocProf	A	20	17	Male	81285
Prof	A	18	18	Male	126300
Prof	A	29	19	Male	94350
Prof	A	51	51	Male	57800
Prof	B	39	33	Male	128250
Prof	B	23	23	Male	134778

Figure 7-1. *Salaries data set*

Direct Plotting

Pandas is a Python library with data frame features that supplies built-in options for plotting visualizations in a two-dimensional tabular style. In Listing 7-1, you read the Salaries data set and create some vectors of variables, which are `rank`, `discipline`, `phd`, `service`, `sex`, and `salary`.

Listing 7-1. Reading the Data Set

```
In [3]: import pandas as pd
        dataset = pd.read_csv("./Data/Salaries.csv")
        rank = dataset['rank']
```

```
discipline = dataset['discipline']
phd = dataset['phd']
service = dataset['service']
sex = dataset['sex']
salary = dataset['salary']

dataset.head()
```

Out[1]:

	rank	discipline	phd	service	sex	salary
0	Prof	B	56	49	Male	186960
1	Prof	A	12	6	Male	93000
2	Prof	A	23	20	Male	110515
3	Prof	A	40	31	Male	131205
4	Prof	B	20	18	Male	104800

Line Plot

You can use line plotting as shown in Listing 7-2. It's important to ensure the data units, such as the phd, service, and salary variables, are used for plotting. However, only the salaries are visible, while the phd and service information is not clearly displayed on the plot. This is because the numerical units in the salaries are in the hundreds of thousands, while the phd and services information is in very small units.

Listing 7-2. Visualizing Patterns with High Differences in Numerical Units

```
In [5]: dataset[["rank", "discipline","phd","service", "sex",
"salary"]].plot()
```

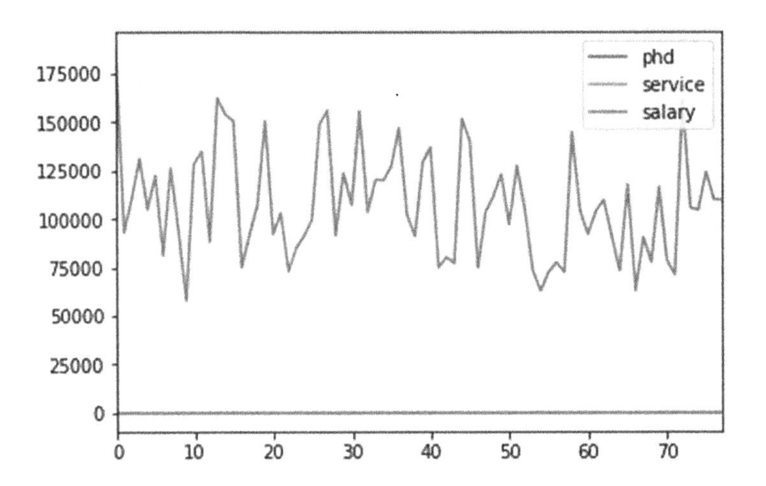

Let's visualize more comparable units such as the `phd` and `services` information, as shown in Listing 7-3. You can observe the correlation between `phd` and `services` over the years, except from age 55 up to 80, where `services` decline, which means that some people left the service at the age of 55 and older.

Listing 7-3. Visualizing Patterns with Close Numerical Units

```
In [6]: dataset[["phd","service"]].plot()
```

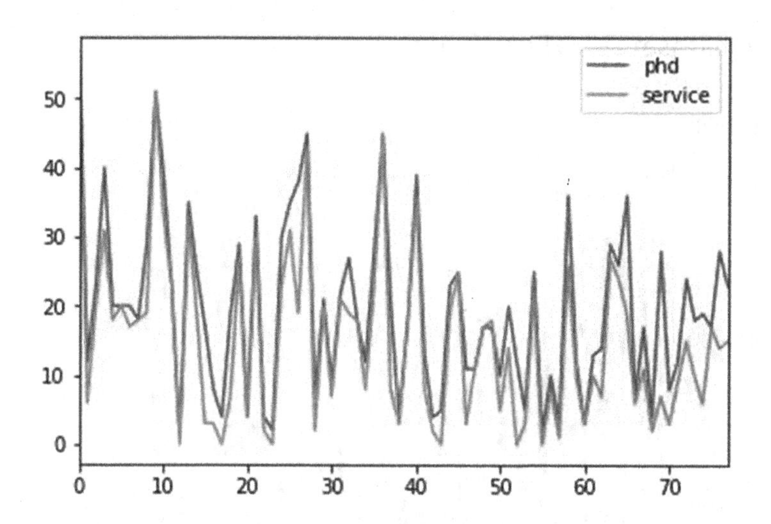

In Listing 7-4, you are grouping data by service and summarizing the salaries per service category. Then you sort the derived data set in descending order according to the salaries. Finally, you plot the sorted data set using a bar chart.

Listing 7-4. Visualizing Salaries per Service Category

```
In [4]: dataset1 = dataset.groupby(['service']).sum()
        dataset1.sort_values("salary", ascending = False,
        inplace=True)
        dataset1.head()
```

Out[4]:

service	phd	salary
19	178	769448
3	56	635216
18	91	603060
0	26	519500
7	70	440408

```
In [8]: dataset1["salary"].plot.bar()
```

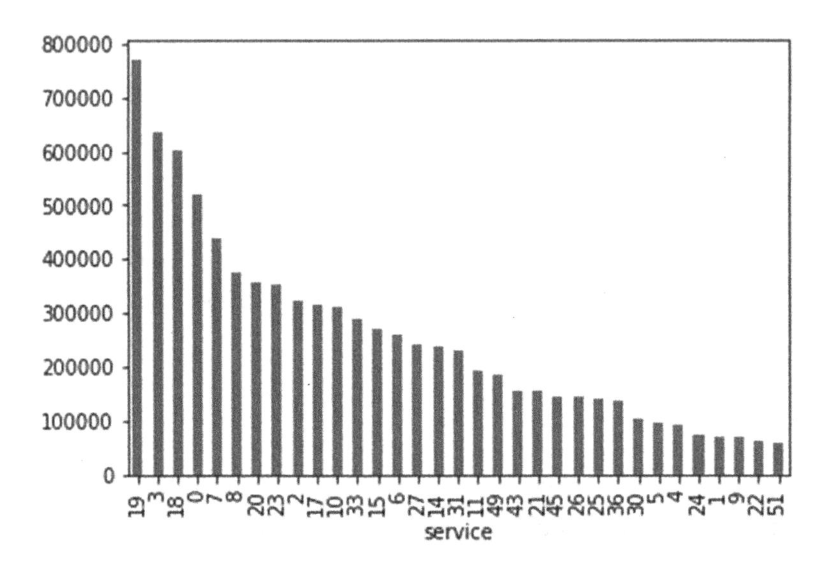

You can see that most people serve approximately 19 years, which is why the highest accumulated salary is from this category.

Bar Plot

Listing 7-5 shows how to plot the first ten records of `phd` and `services`, and you can add a title as well. To add a title to the chart, you need to use `.bar(title="Your title")`.

Listing 7-5. Bar Plotting

```
In [9]: dataset[[ 'phd', 'service' ]].head(10).plot.bar()
```

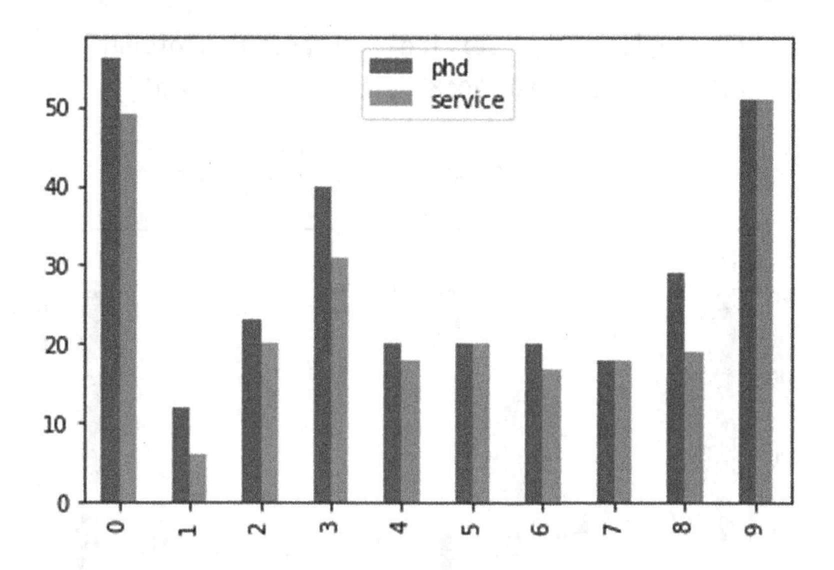


```
In [11]: dataset[['phd', 'service']].head(10).plot.bar
(title="Ph.D. Vs Service\n 2018")
```

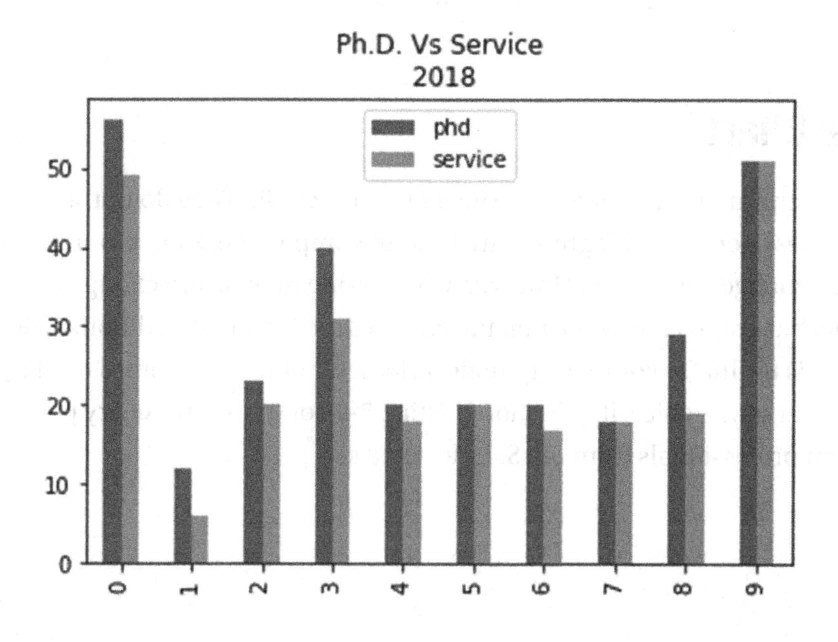

```
In [12]: dataset[['phd', 'service']].head(10).plot.bar
(title="Ph.D. Vs Service\n 2018" , color=['g','red'])
```

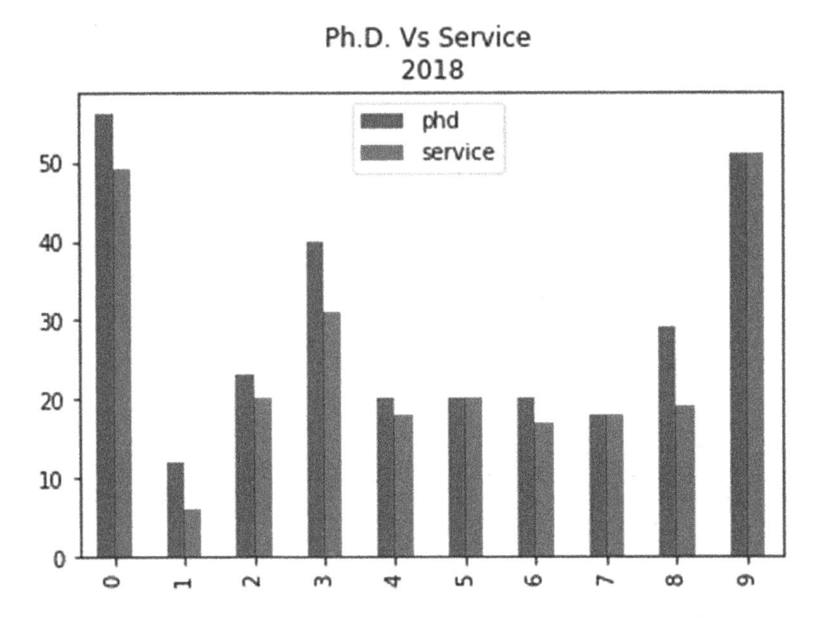

Pie Chart

Pie charts are useful for comparing parts of a whole. They do not show changes over time. Bar graphs are used to compare different groups or to track changes over time. However, when trying to measure change over time, bar graphs are best when the changes are larger. In addition, a pie chart is useful for comparing small variables, but when it comes to a large number of variables, it falls short. Listing 7-6 compares the salary package of ten professionals from the Salaries data set.

Listing 7-6. Pie Chart

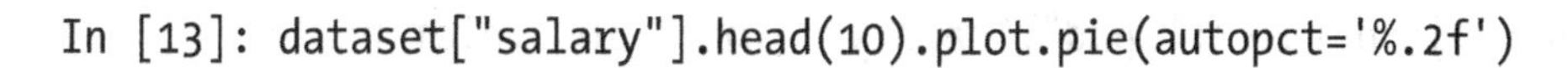

```
In [13]: dataset["salary"].head(10).plot.pie(autopct='%.2f')
```

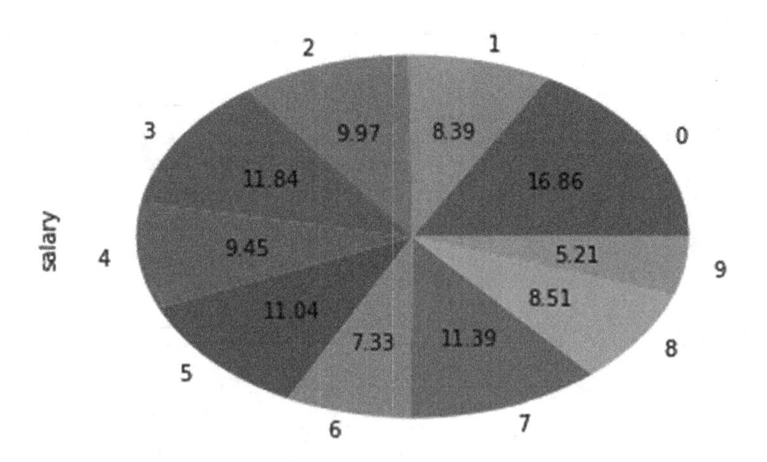

Box Plot

Box plotting is used to compare variables using some statistical values. The comparable variables should be of the same data units; Listing 7-7 shows that when you compare phd and `salary`, it produces improper figures and does not provide real comparison information since the `salary` numerical units are much higher than the phd numerical values. Plotting phd and `services` shows that the median and quantiles of phd are higher than the median and quantiles of the `service` information; in addition, the range of phd is wider than the range of `service` information.

Listing 7-7. Box Plotting

```
In [14]: dataset[["phd","salary"]].head(100).plot.box()
```

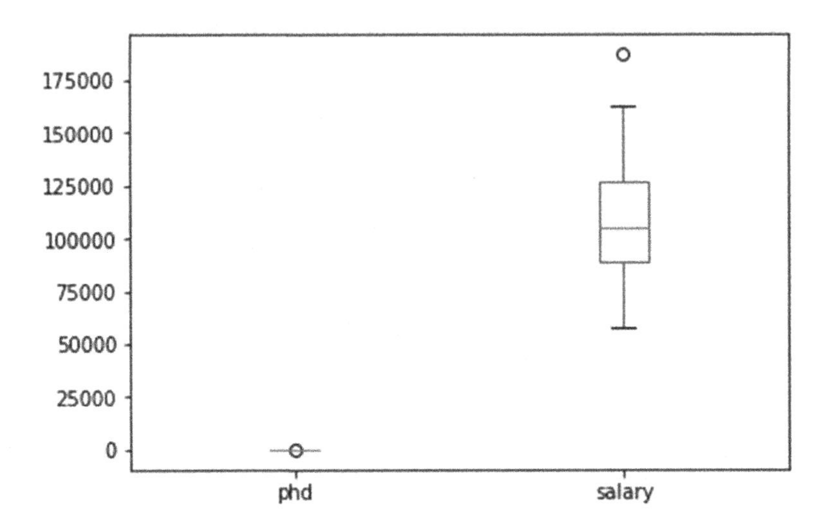


```
In [15]: dataset[["phd","service"]].plot.box()
```

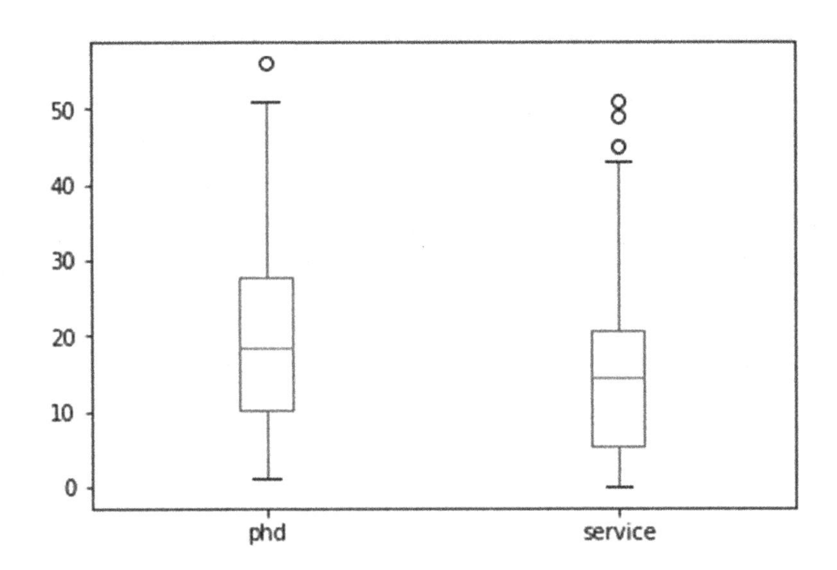

Histogram Plot

A histogram can be used to represent a specific variable or set of variables. Listing 7-8 plots 20 records of the `salaries` variables; it shows that salary packages of about 135,000 are the most frequent in this data set.

Listing 7-8. Histogram Plotting

```
In [16]: dataset["salary"].head(20).plot.hist()
```

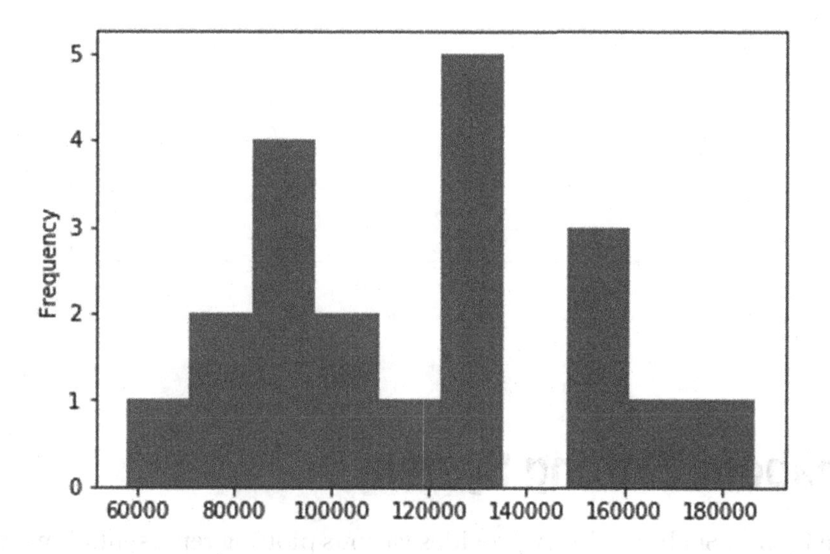

Scatter Plot

A scatter plot shows the relationship between two factors of an experiment (e.g. phd and service). A trend line is used to determine positive, negative, or no correlation. See Listing 7-9.

Listing 7-9. Scatter Plotting

```
In [17]: dataset.plot(kind='scatter', x='phd', y='service',
title='Popuation vs area and density\n 2018', s=0.9)
```

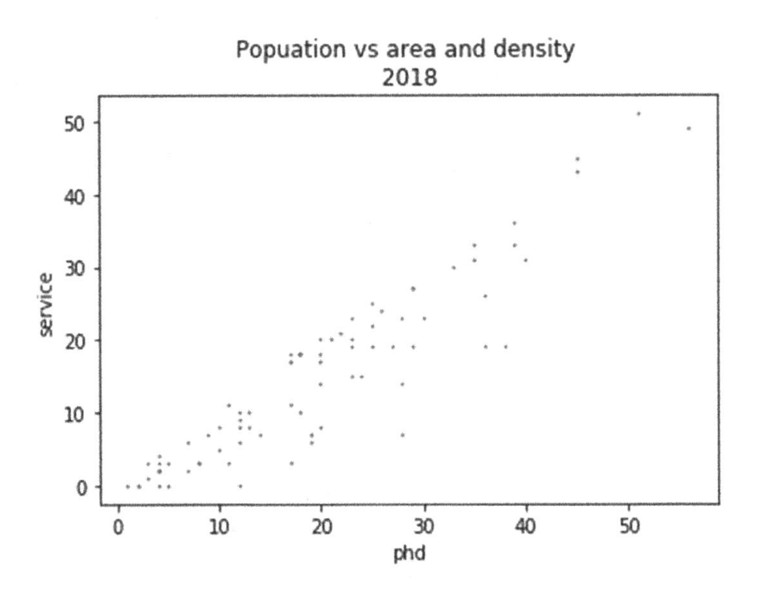

Seaborn Plotting System

The Python Seaborn library provides various plotting representations for visualizing data. A strip plot is a scatter plot where one of the variables is categorical. Strip plots can be combined with other plots to provide additional information. For example, a box plot with an overlaid strip plot is similar to a violin plot because some additional information about how the underlying data is distributed becomes visible. Seaborn's swarm plot is virtually identical to a strip plot except that it prevents data points from overlapping.

Strip Plot

Listing 7-10 uses strip plotting to display data per salary category.

Listing 7-10. Simple Strip Plot

```
In [3]: # Simple stripplot sns.stripplot( x =
dataset['salary'])
```

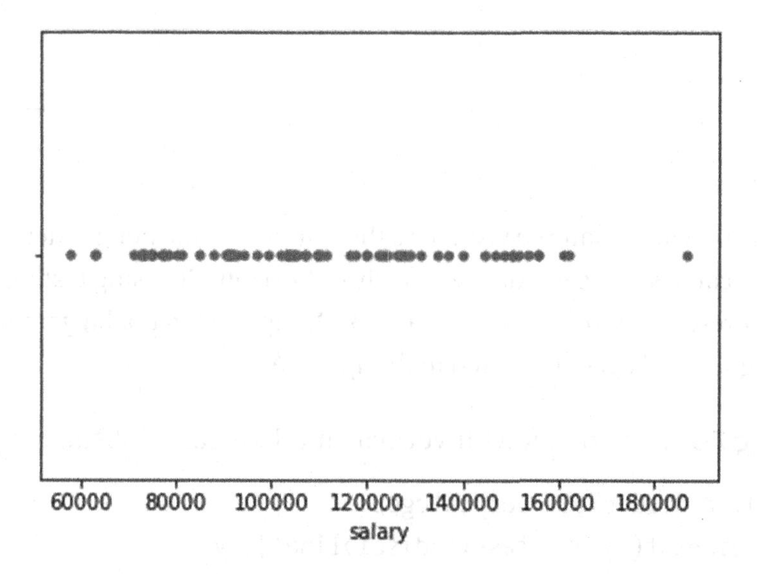


```
In [4]: # Stripplot over categories
sns.stripplot( x = dataset['sex'], y= dataset['salary'],
data=dataset);
```

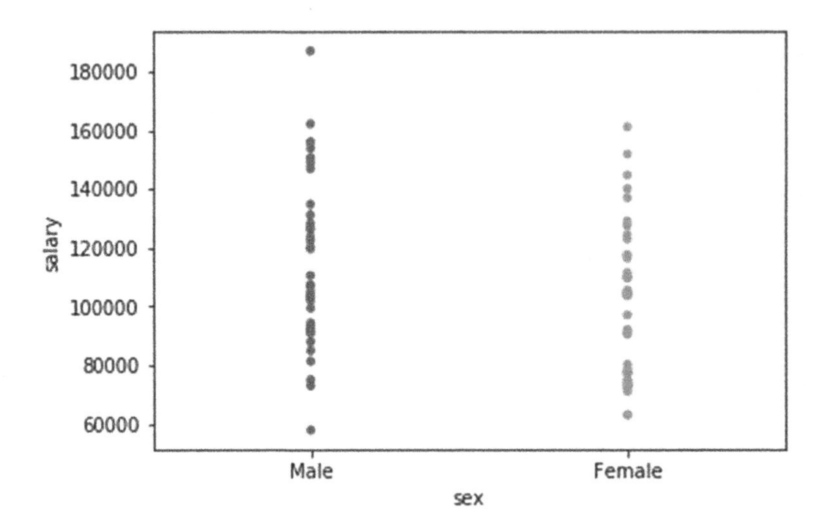

The previous example visualizes the salary variable per gender.

You can visualize the data vertically or horizontally using Listing 7-11, which presents two disciplines, A and B. Discipline B has a bigger range and higher packages compared to discipline A.

Listing 7-11. Strip Plot with Vertical and Horizontal Visualizing

```
In [5]: # Stripplot over categories
sns.stripplot( x = dataset['discipline'], y =
dataset['salary'], data=dataset, jitter=1)
```

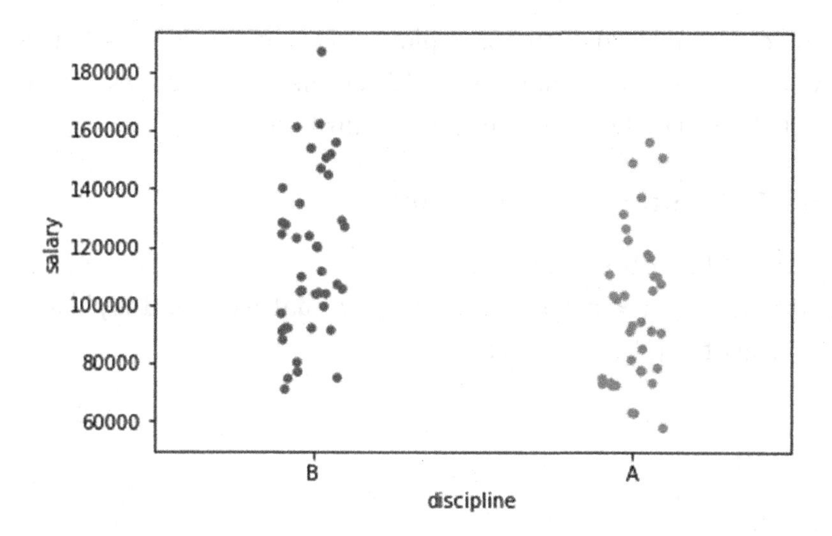


```
In [6]: # Stripplot over categories Horizontal
sns.stripplot( x= dataset['salary'], y = dataset['discipline'],
data=dataset, jitter=True);
```

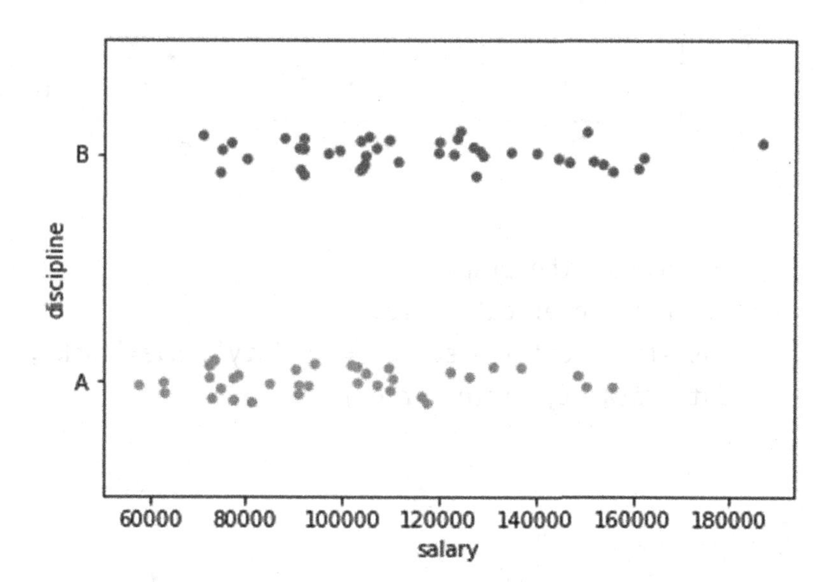

You can visualize data in a strip plot per category; Listing 7-12 uses the assistance prof, associate prof, and full professor categories. The hue attribute is used to determine the legend attribute.

Listing 7-12. Strip Plot per Category

```
In [7]: # Stripplot over categories
sns.stripplot( x = dataset['rank'], y= dataset['salary'],
data=dataset, jitter=True);
```

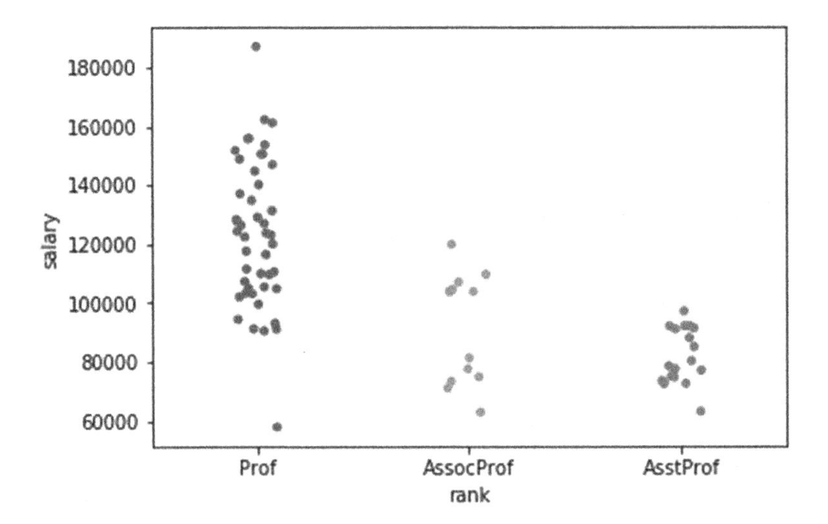


```
In [8]: # Add hue to the graph
        # Stripplot over categories
          sns.stripplot( x ='sex', y= 'salary', hue='rank',
          data=dataset, jitter=True )
```

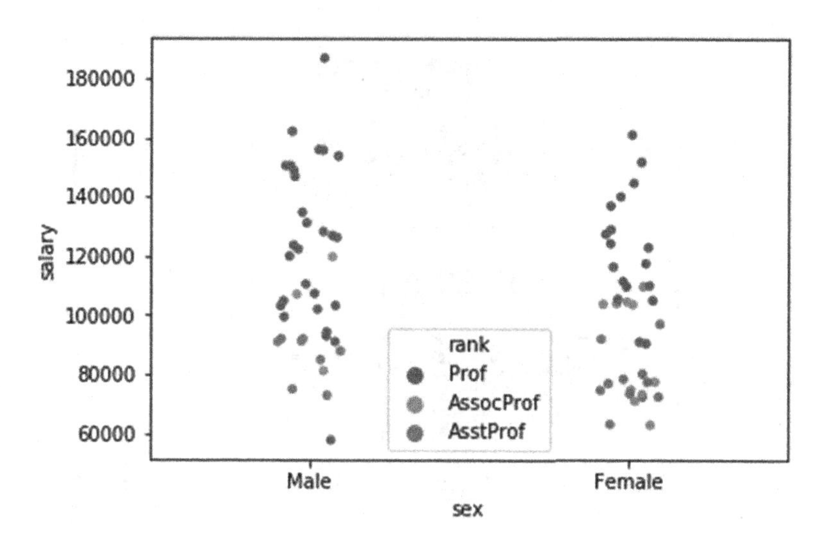

Box Plot

You can combine a box plot and strip plot to give more information on the generated plot (see Listing 7-13). As shown, the Male category has a higher median salary, maximum salary, and range compared to the Female category.

Listing 7-13. Combined Box Plot and Strip Plot Visualization

```
In [10]: # Draw data on top of boxplot
         sns.boxplot(x = 'salary', y ='sex', data=dataset,
         whis=np.inf )
         sns.stripplot(x = 'salary', y ='sex', data=dataset,
         jitter=True, color='0.02')
```

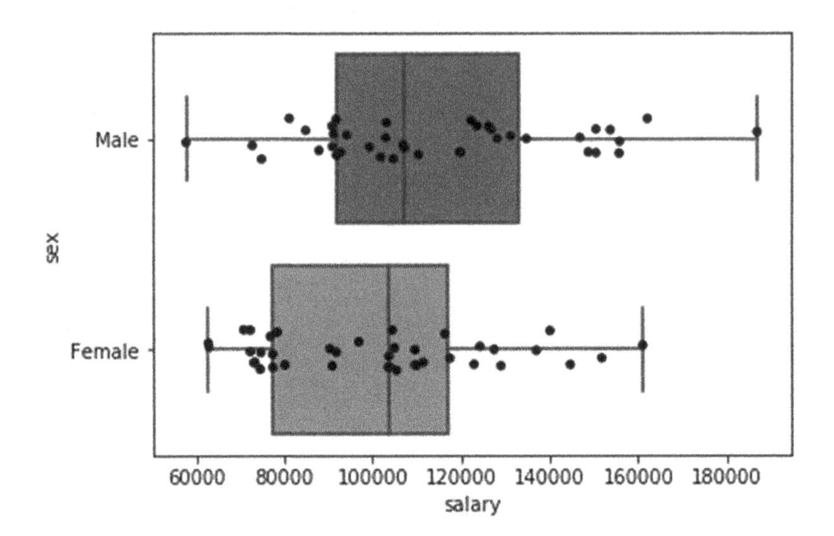


```
In [13]: # box plot salaries
         sns.boxplot(x = dataset['salary'])
```

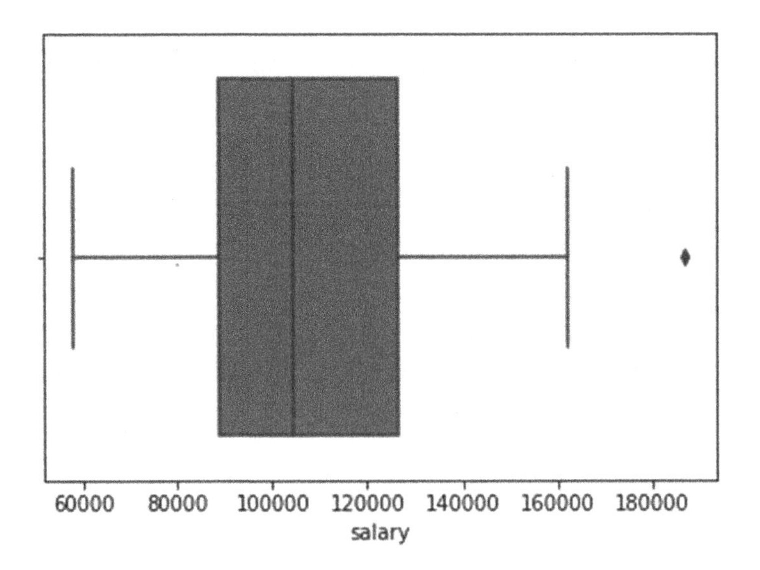

```
In [14]: # box plot salaries
         sns.boxplot(x = dataset['salary'], notch=True)
```

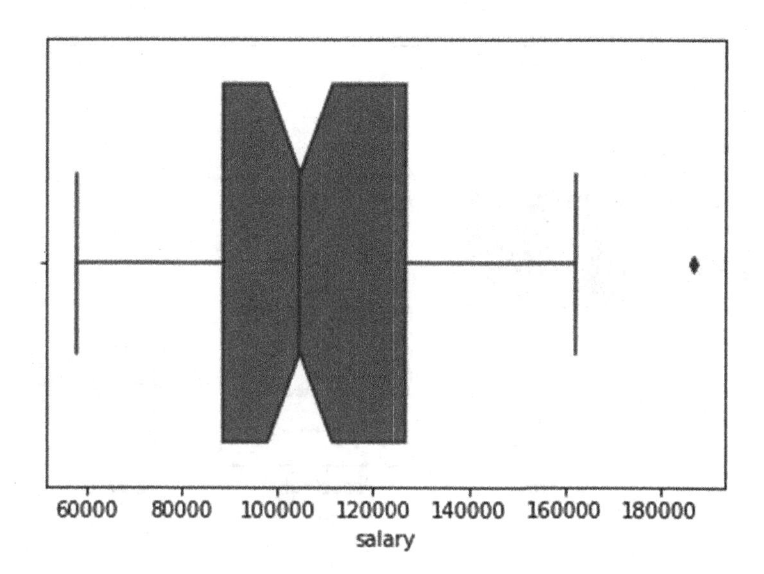


```
In [15]: # box plot salaries
         sns.boxplot(x = dataset['salary'], whis=2)
```

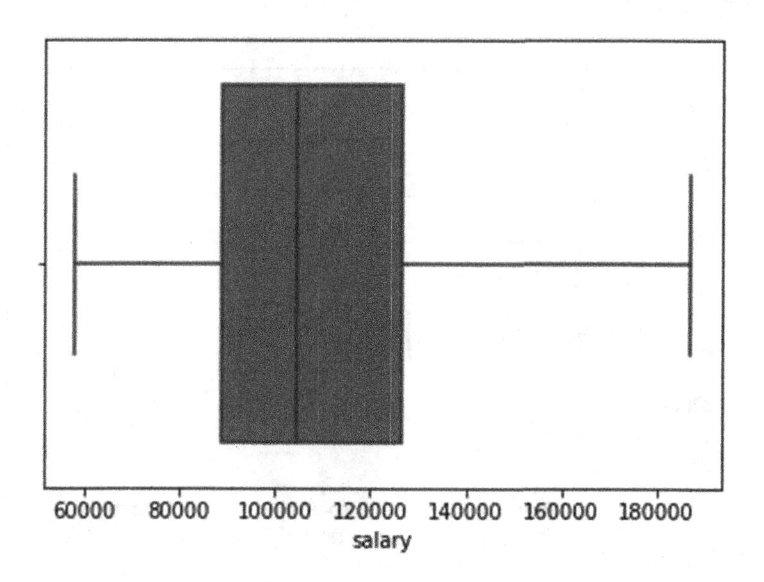

```
In [16]: # box plot per rank
         sns.boxplot(x = 'rank', y = 'salary', data=dataset)
```

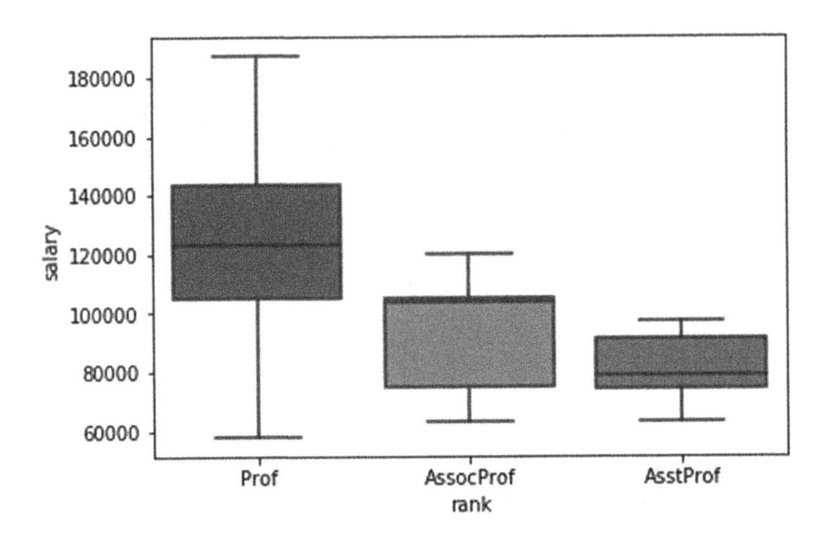


```
In [17]: # box plot per rank
sns.boxplot(x = 'rank', y = 'salary', hue='sex', data=dataset,
palette='Set3')
```

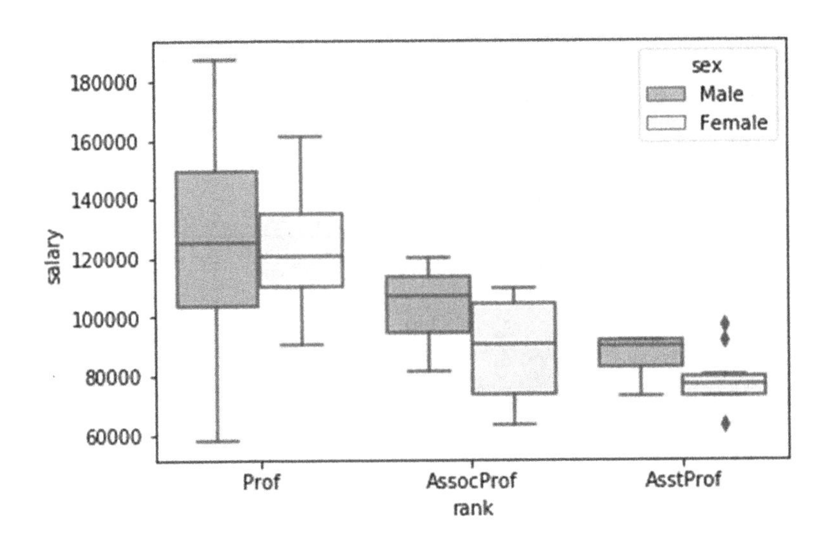

```
In [18]: # box plot per rank
         sns.boxplot(x = 'rank', y = 'salary', data=dataset)
         sns.swarmplot(x = 'rank', y = 'salary', data=dataset,
         color='0.25')
```

Combined Box Plot and Strip Plot Visualization as shown in below figure.

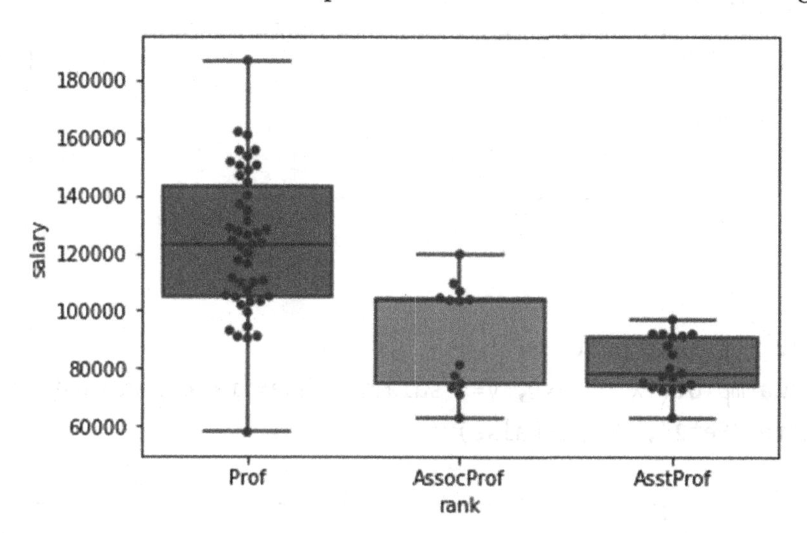

Swarm Plot

A swarm plot is used to visualize different categories; it gives a clear picture of a variable distribution against other variables. For instance, the salary distribution per gender and per profession indicates that the male professors have the highest salary range. Most of the males are full professors, then associate, and then assistant professors. There are more male professors than female professors, but there are more female associate professors than male associate professors. See Listing 7-14.

Listing 7-14. Swarm ploting of salary against gender

```
In [11]: # swarmplot
sns.swarmplot( x ='sex', y= 'salary', hue='rank', data=dataset,
palette="Set2", dodge=True)
```

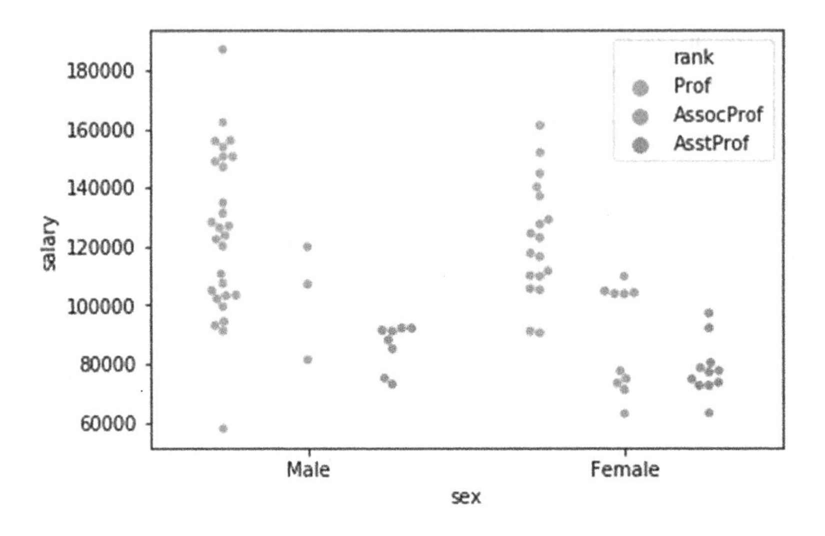


```
In [12]: # swarmplot
sns.swarmplot( x ='sex', y= 'salary', hue='rank', data=dataset,
palette="Set2", dodge=False)
```

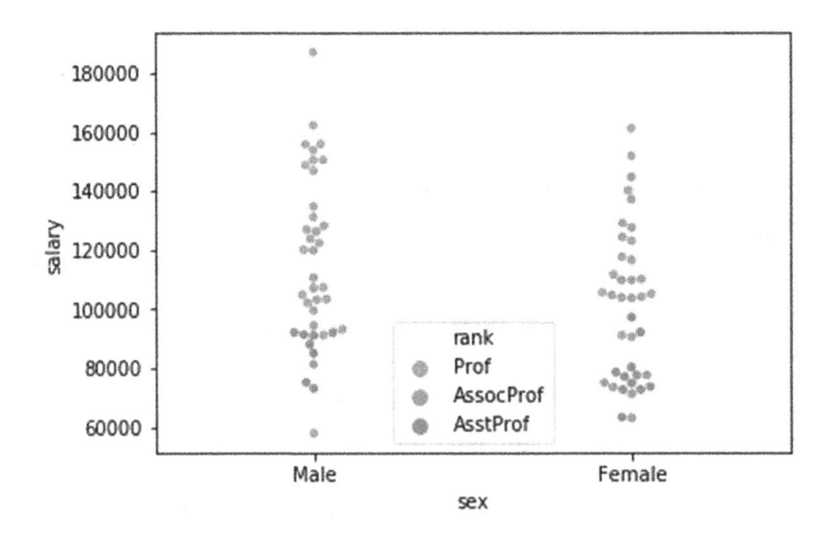

Joint Plot

A joint plot combines more than one plot to visualize the selected patterns (see Listing 7-15).

Listing 7-15. Joint Plot Visualization

```
In [22]: sns.jointplot(x = 'salary', y = 'service',
data=dataset)
```

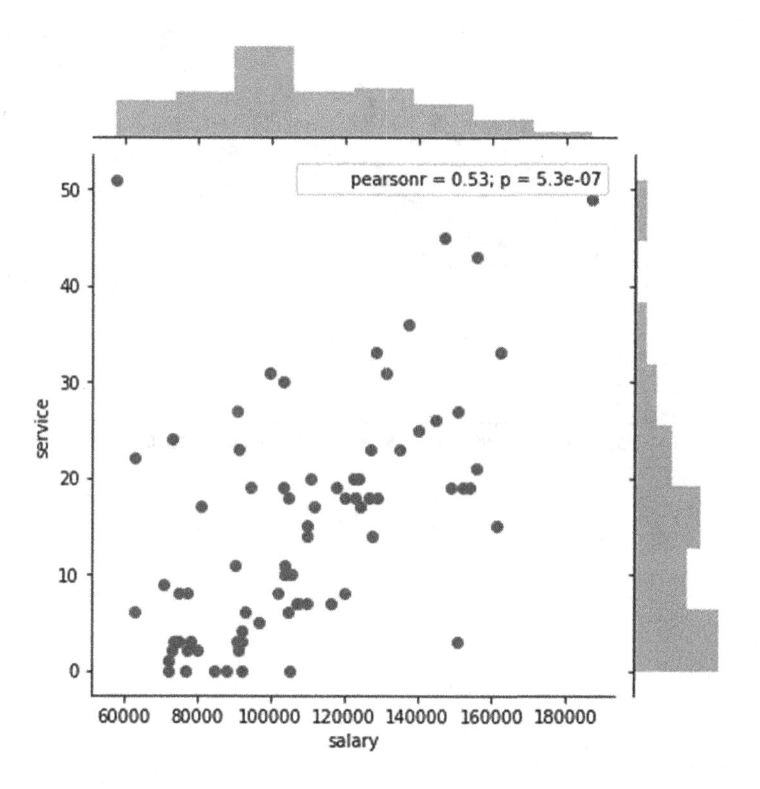

```
In [24]: sns.jointplot('salary', 'service', data=dataset,
kind='reg')
```

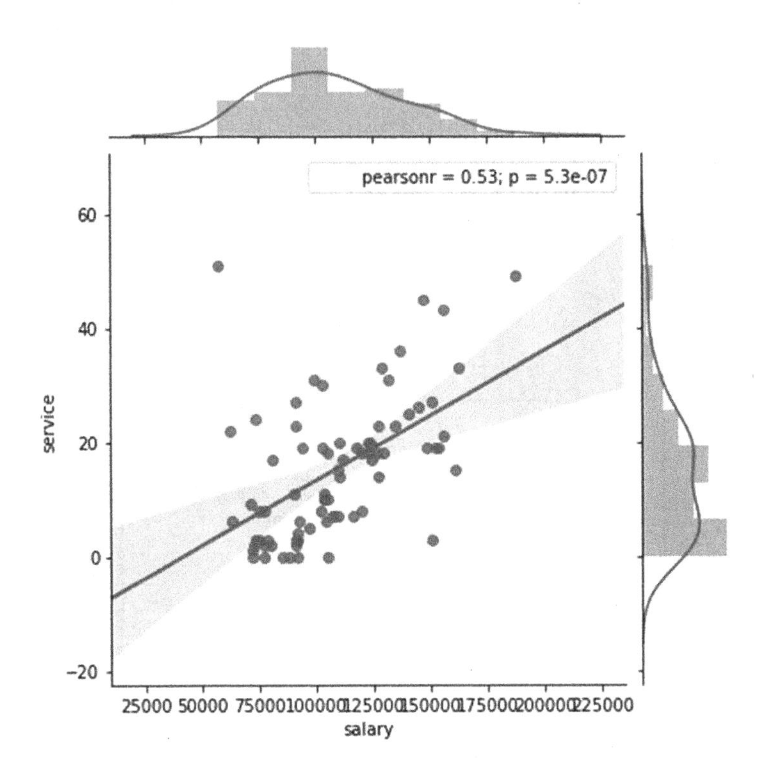

```
In [25]: sns.jointplot('salary', 'service', data=dataset,
kind='hex')
```

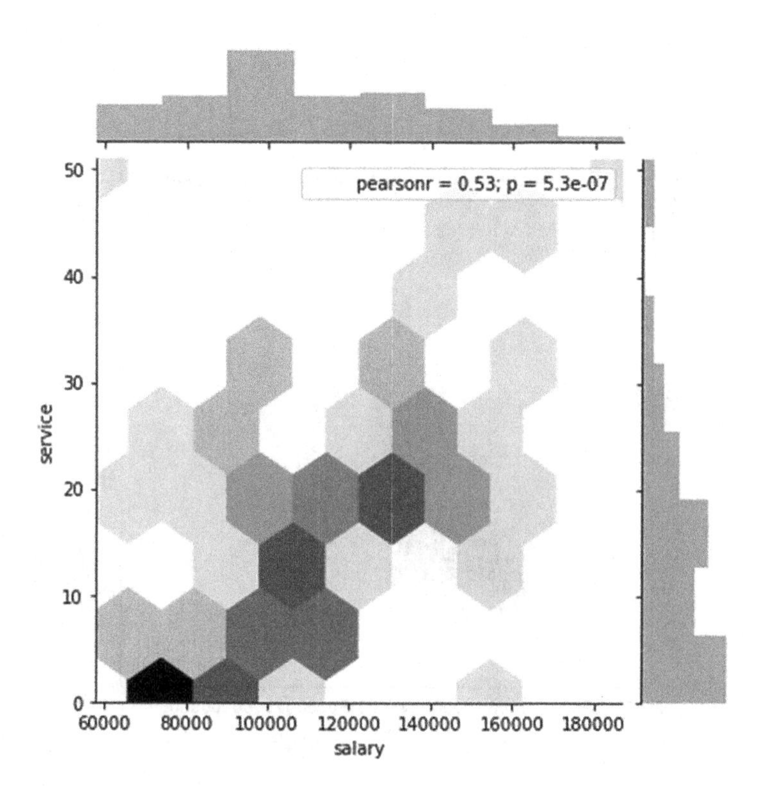

```
In [26]: sns.jointplot('salary', 'service', data=dataset,
kind='kde')
```

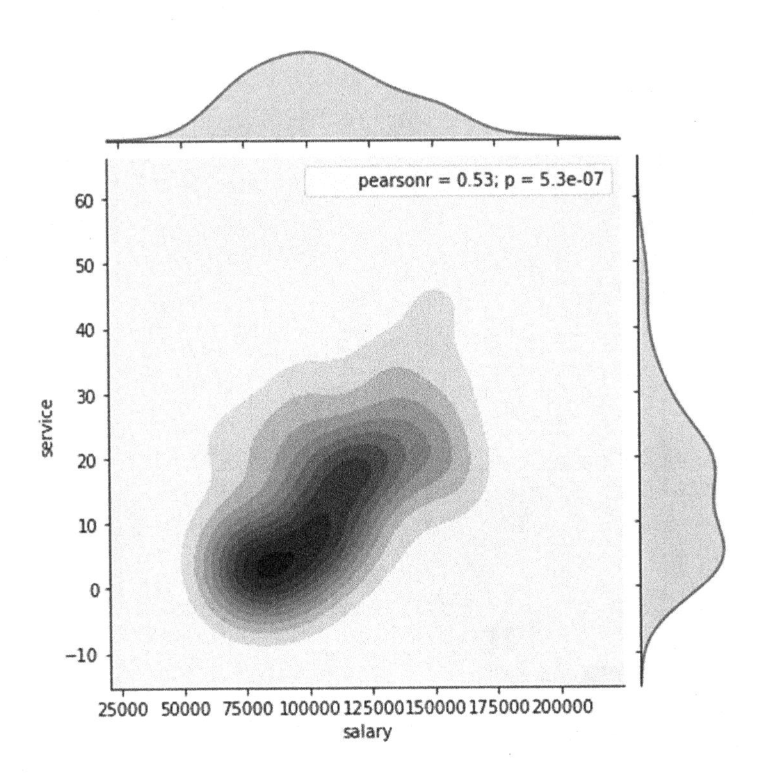

```
In [27]: from scipy.stats import spearmanr sns.
jointplot('salary', 'service', data=dataset, stat_func=
spearmanr )
```

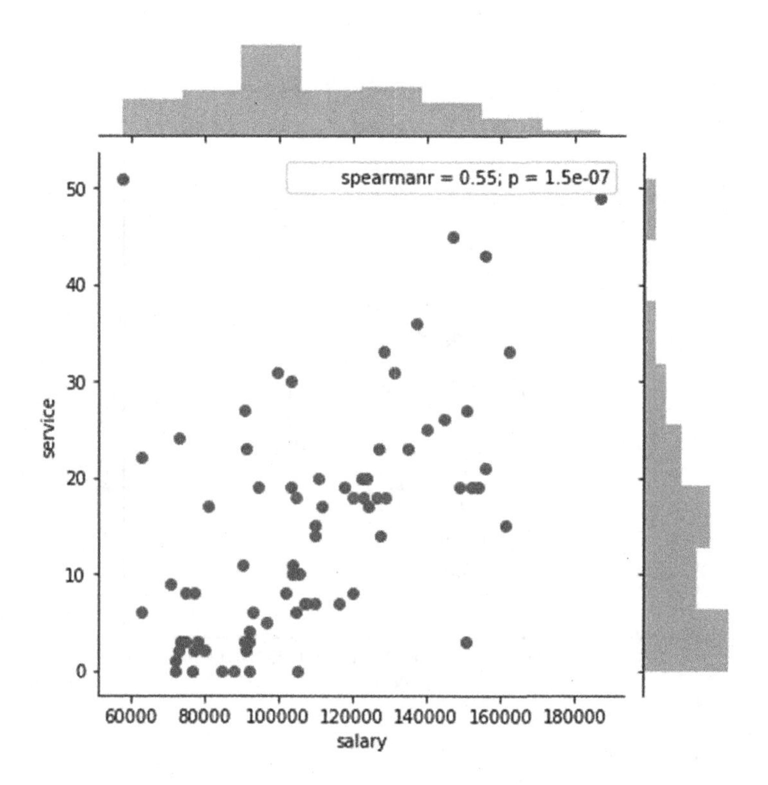

```
In [31]: sns.jointplot('salary', 'service',
         data=dataset).plot_joint(sns.kdeplot, n_levels=6)
```

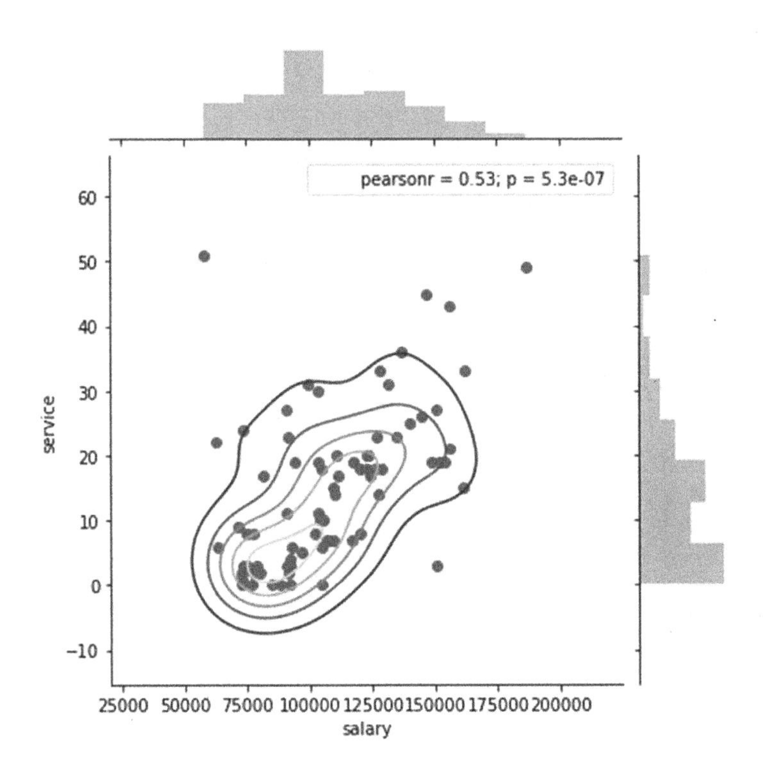

```
In [32]: sns.jointplot('salary', 'service',
         data=dataset).plot_joint( sns.kdeplot,n_levels=6).
         plot_marginals(sns.rugplot)
```

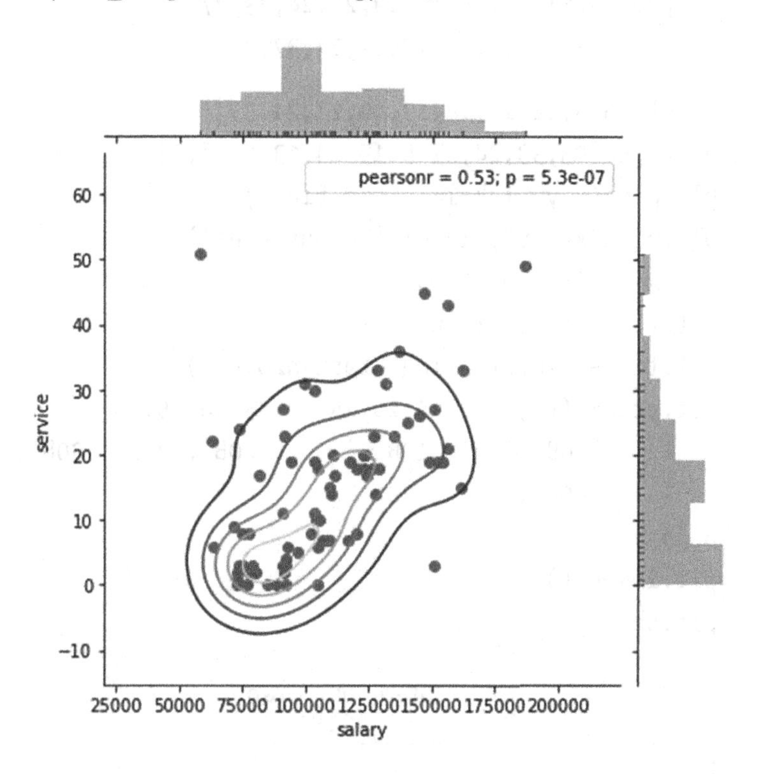

Matplotlib Plot

Matplotlib is a Python 2D plotting library that produces high-quality figures in a variety of hard-copy formats and interactive environments across platforms. In Matplotlib, you can add features one by one, such as adding a title, labels, legends, and more.

Line Plot

In inline plotting, you should determine the x- and y-axes, and then you can add more features such as a title, a legend, and more (see Listing 7-16).

Listing 7-16. Matplotlib Line Plotting

```
In [2]: import matplotlib.pyplot as plt
        x =[3,6,8,11,13,14,17,19,21,24,33,37]
        y = [7.5,12,13.2,15,17,22,24,37,34,38.5,42,47]

        x2 =[3,6,8,11,13,14,17,19,21,24,33]
        y2 = [50,45,33,24,21.5,19,14,13,10,6,3]
        plt.plot(x,y, label='First Line')
        plt.plot(x2, y2, label='Second Line')
        plt.xlabel('Plot Number')
        plt.ylabel('Important var')
        plt.title('Interesting Graph\n2018 ')
        plt.yticks([0,5,10,15,20,25,30,35,40,45,50],
                ['0B','5B','10B','15B','20B','25B','30B','35B',
                '40B','45B','50
        B'])
        plt.legend()
        plt.show()
```

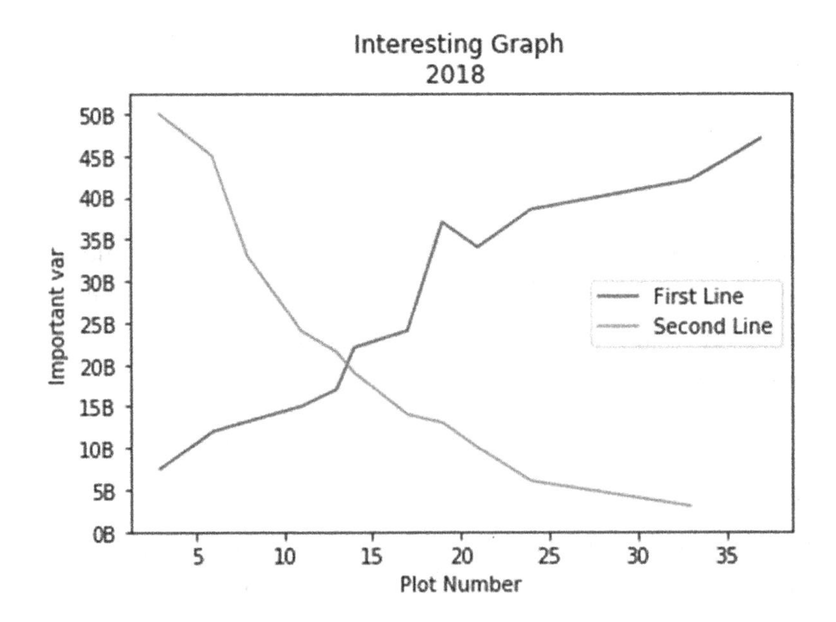

```
In [13]: plt.plot(phd, label='Ph.D.')
         plt.plot(service, label='service')
         plt.xlabel('Ph.D./service')
         plt.ylabel('Frequency')
         plt.title('Ph.D./service\nDistribution')
         plt.legend()
         plt.show()
```

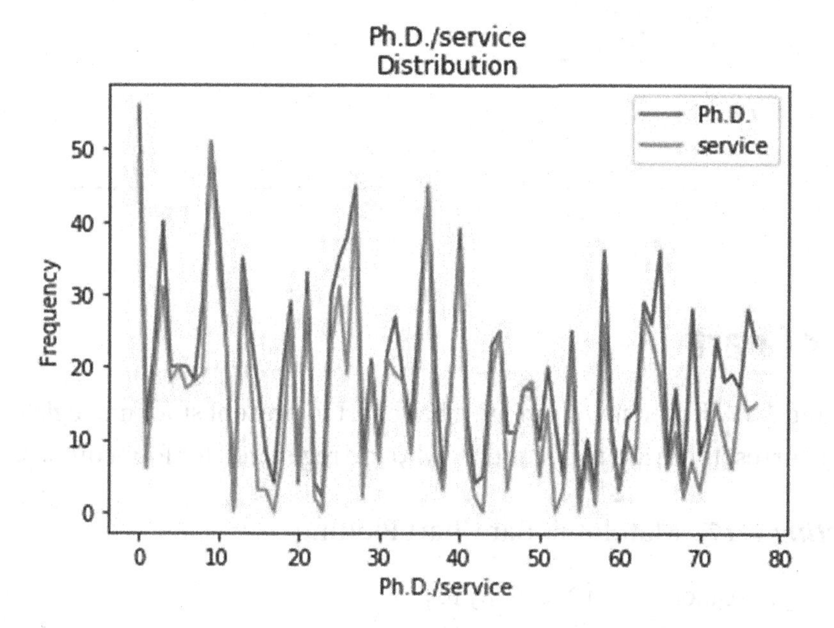


```
In [15]: plt.plot(phd, service, 'bo', label="Ph.D. Vs
services", lw=10)
         plt.grid()
         plt.legend()
         plt.xlabel('Ph.D')
         plt.ylabel('service')
         plt.title('Ph.D./salary\nDistribution')
         plt.yscale('log')
```

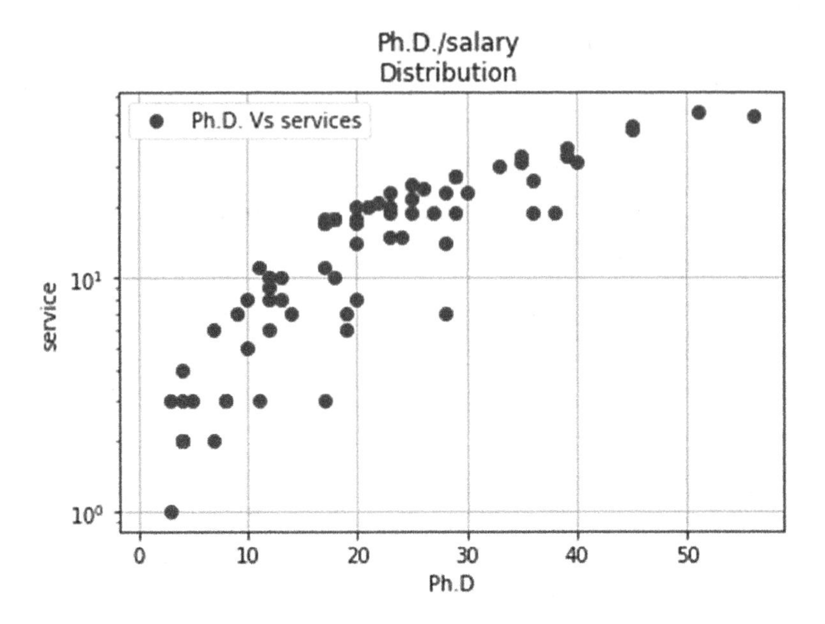

Bar Chart

Listing 7-17 shows how to create a bar chart to present students registered for courses; there are two students who are registered for four courses.

Listing 7-17. Matplotlib Bar Chart Plotting

```
In [3]: Students = [2,4,6,8,10]
        Courses = [4,5,3,2,1]
        plt.bar(Students,Courses, label="Students/Courses")
        plt.xlabel('Students ')
        plt.ylabel('Courses')
        plt.title('Students Courses Data\n 2018')
        plt.legend()
        plt.show()
```

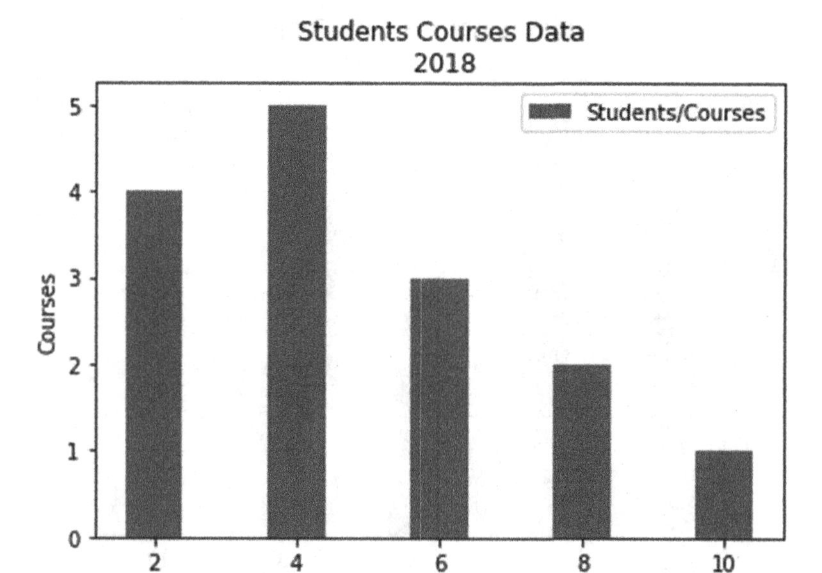


```
In [4]: Students = [2,4,6,8,10]
        Courses = [4,5,3,2,3]
        stds = [3,5,7,9,11]
        Projects = [1,2,4,3,2]
        plt.bar(Students, Courses, label="Courses", color='r')
        plt.bar(stds, Projects, label="Projects", color='c')
        plt.xlabel('Students')
        plt.ylabel('Courses/Projects')
        plt.title('Students Courses and Projects Data\n 2018')
        plt.legend()
        plt.show()
```

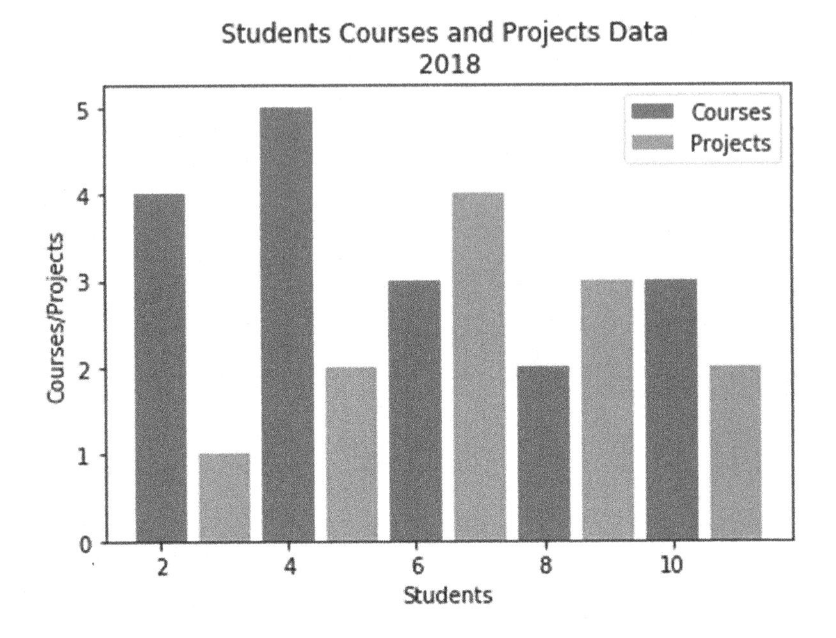

Histogram Plot

Listing 7-18 shows how to create a histogram showing age frequencies; most people in the data set are between 30 and 40. In addition, you can create a histogram of the years of service and the number of PhDs.

Listing 7-18. Matplotlib Histogram Plotting

```
In [5]: Ages = [22.5, 10, 55, 8, 62, 45, 21, 34, 42, 45, 99,
              75, 82,
               77, 55, 43, 66, 66, 78, 89, 101, 34, 65, 56,
               25, 34,
               52, 25, 63, 37, 32]
        binsx = [0, 10, 20, 30, 40, 50, 60, 70, 80, 90, 100, 110]
        plt.hist(Ages, bins=binsx, histtype='bar', rwidth=0.7)
```

```
plt.xlabel('Ages')
plt.ylabel('Frequency')
plt.title('Ages frequency for sample pouplation\n 2018')
plt.show()
```

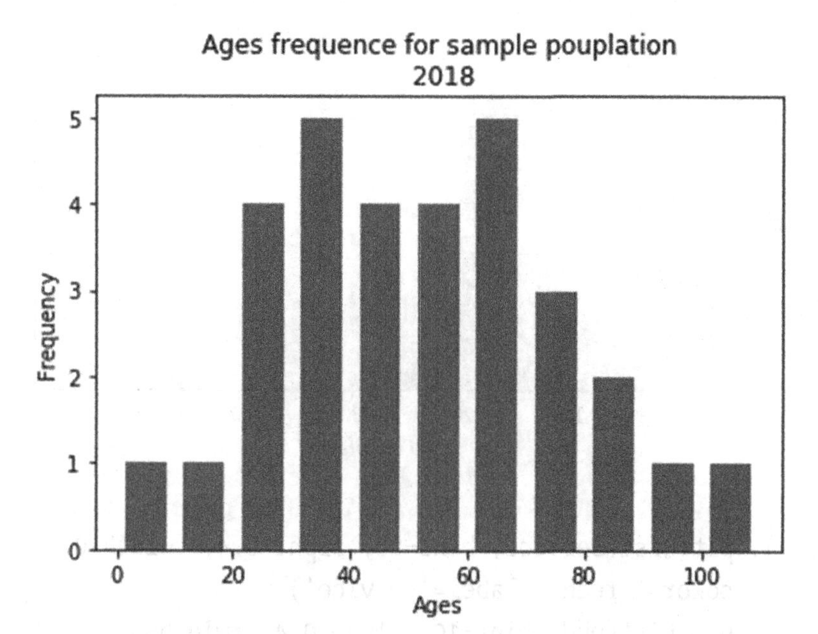


```
In [18]: plt.hist(service, bins=30, alpha=0.4, rwidth=0.8,
         color='green', label='service')
         plt.hist(phd, bins=30, alpha=0.4, rwidth=0.8,
         color='red', label='phd')
         plt.xlabel('Services/phd')
         plt.ylabel('Distribution')
         plt.title('Services/phd\n 2018')
         plt.legend(loc='upper right')
         plt.show()
```

Visualize service years since Ph.D. had attained.

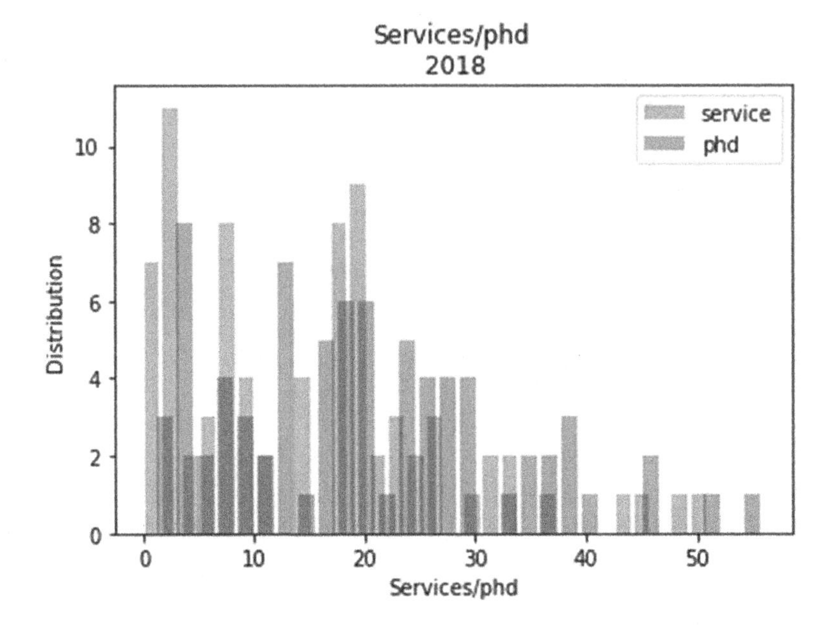


```
In [19]: plt.hist(service, bins=10, alpha=0.4, rwidth=0.8,
         color='green', label='service')
         plt.hist(phd, bins=10, alpha=0.4, rwidth=0.8,
         color='red', label='phd')
         plt.xlabel('Services/phd')
         plt.ylabel('Distribution')
         plt.title('Services/phd\n 2018')
         plt.legend(loc='upper right')
         plt.show()
```

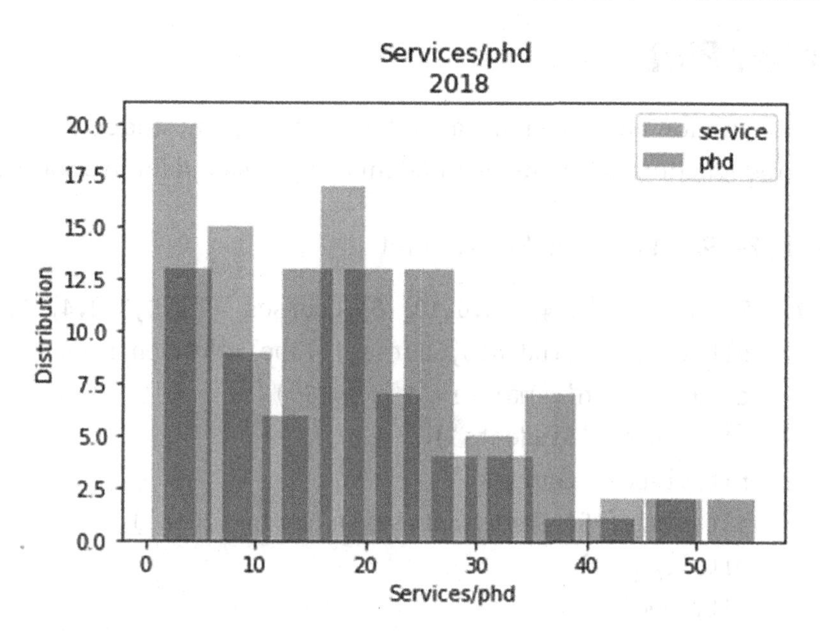


```
In [21]: plt.hist(salary, bins=100)
         plt.show()
```

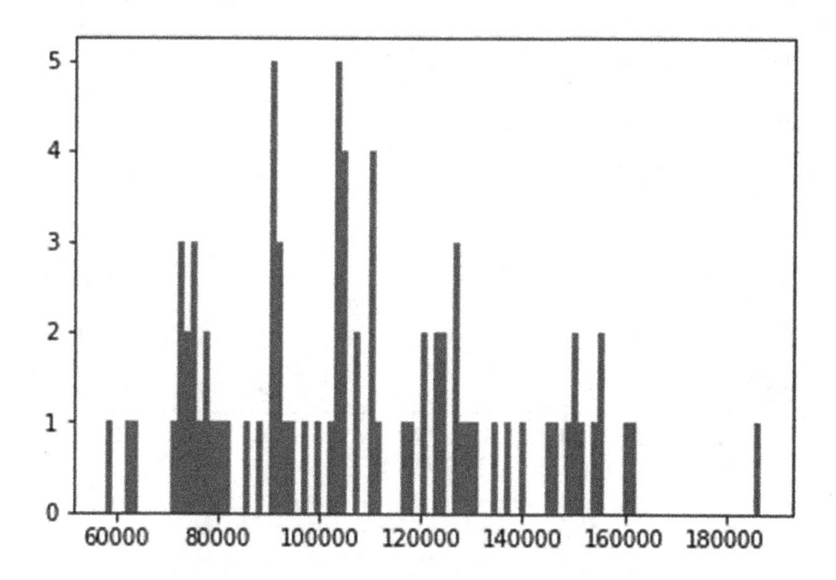

Scatter Plot

Listing 7-19 shows how to create a scatter plot to present students registered for courses, where four students are registered for five courses.

Listing 7-19. Matplotlib Scatter Plot

```
In [7]: Students = [2,4,6,8,6,10, 6] Courses = [4,5,3,2,4, 3, 4]
        plt.scatter(Students,Courses, label='Students/Courses',
        color='green', marker='*', s=75 )
        plt.xlabel('Students')
        plt.ylabel('Courses')
        plt.title('Students courses\n Spring 2018')
        plt.legend()
        plt.show()
```

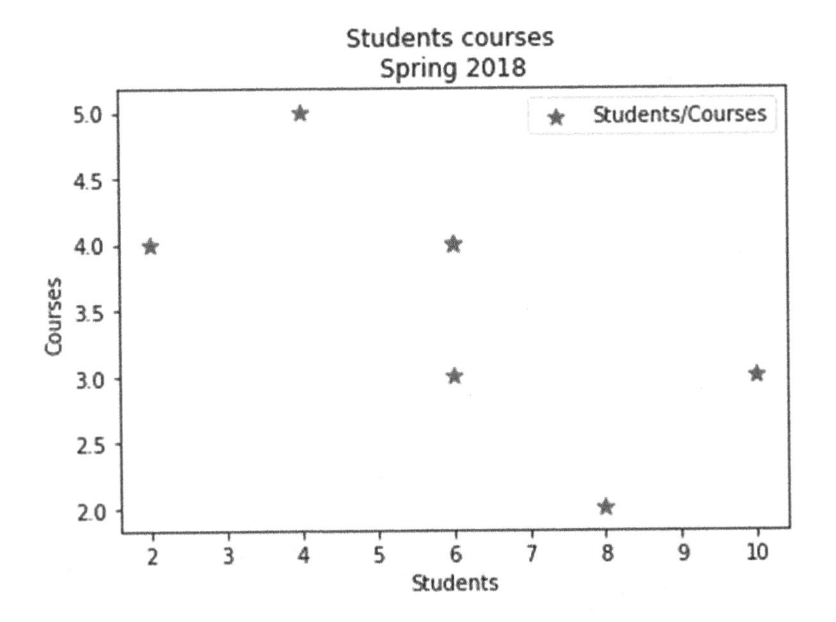

```
In [16]: plt.scatter(rank,salary, label='salary/rank',
         color='g', marker='+', s=50 )
         plt.xlabel('rank') plt.ylabel('salary')
         plt.title('salary/rank\n Spring 2018')
         plt.legend() plt.show()
```

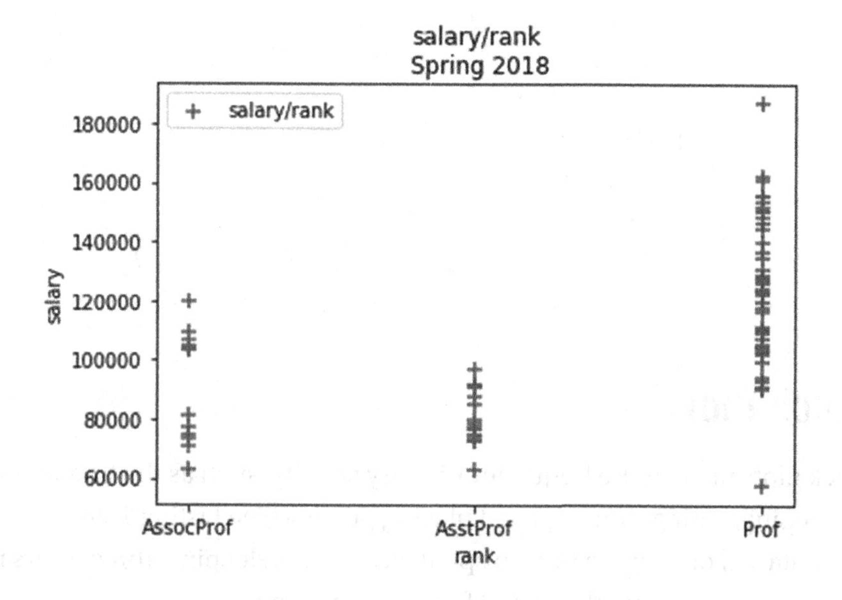


```
In [20]: plt.scatter(phd,salary, label='Salary/phd', color='g',
         marker='+', s=80 )
         plt.xlabel('phd') plt.ylabel('salary')
         plt.title('phd/ salary\n Spring 2018')
         plt.legend() plt.show()
```

Stack Plot

Stack plots present the frequency of every activity, such as the frequency of sleeping, eating, working, and playing per day (see Listing 7-20). In this data set, on day 2, a person spent eight hours sleeping, three hours in eating, eight hours working, and five hours playing.

Listing 7-20. Persons Weekly Spent Time per activities using Matplotlib Stack Plot

```
In [9]: days = [1,2,3,4,5]
        sleeping = [7,8,6,11,7]
        eating = [2,3,4,3,2]
        working = [7,8,7,2,2]
        playing = [8,5,7,8,13]
        plt.plot([],[], color='m', label='Sleeping')
        plt.plot([],[], color='c', label='Eating')
        plt.plot([],[], color='r', label='Working')
```

```
plt.plot([],[], color='k', label='Playing')
plt.stackplot(days, sleeping, eating, working ,
playing, colors=['m','c', 'r', 'k'])
plt.xlabel('days')
plt.ylabel('Activities')
plt.title('Persons Weekly Spent Time per activities\n
Spring 2018')
plt.legend()
plt.show()
```

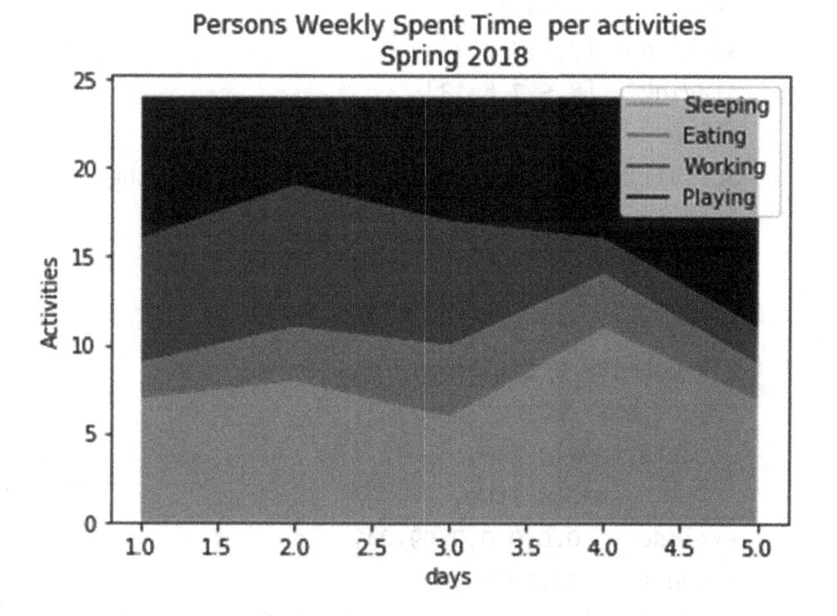

Pie Chart

In Listing 7-21, you are using the explode attribute to slice out a specific activity. After that, you can add the gender and title to the pie chart.

Listing 7-21. Persons Weekly Spent Time per activities using Matplotlib Pie Chart

```
In [10]: days = [1,2,3,4,5]
         sleeping = [7,8,6,11,7]
         eating = [2,3,4,3,2]
         working = [7,8,7,2,2]
         playing = [8,5,7,8,13]
         slices = [39,14,26,41]
         activities = ['sleeping', 'eating', 'working',
         'playing']
         cols = ['c','m','r', 'b','g']

         plt.pie(slices,
             labels= activities,
             colors= cols,
             startangle=100,
                 shadow=True,
         explode = (0.0,0.0,0.09,0),
         autopct = '%1.1f%%')
         plt.title('Persons Weekly Spent Time per activities\n
         Spring 2018')
         plt.legend()
         plt.show()
```

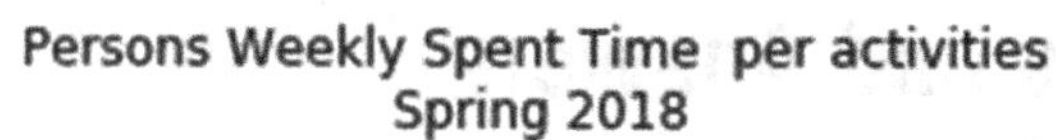

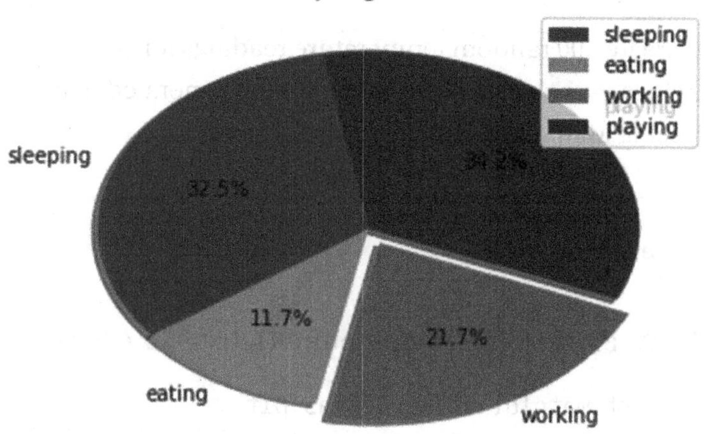

Summary

This chapter covered how to plot data from different collection structures. You learned the following:

- How to directly plot data from a series, data frame, or panel using Python plotting tools such as line plots, bar plots, pie charts, box plots, histogram plots, and scatter plots
- How to implement the Seaborn plotting system using strip plotting, box plotting, swarm plotting, and joint plotting
- How to implement Matplotlib plotting using line plots, bar charts, histogram plots, scatter plots, stack plots, and pie charts

The next chapter will cover the techniques you've studied in this book via two different case studies; it will make recommendations, and much more.

Exercises and Answers

1. Create 500 random temperature readings for six cities over a season and then plot the generated data using Matplotlib.

Answer:

See Listing 7-22.

Listing 7-22. Plotting the Temperature Data of Six Cities

```
In [4]: import matplotlib.pyplot as plt
        plt.style.use('classic')
        %matplotlib inline
        import numpy as np
        import pandas as pd

In [30]: # Create temperature data
        rng = np.random.RandomState(0)
        season1 = np.cumsum(rng.randn(500, 6), 0)

In [32]: # Plot the data with Matplotlib defaults
        plt.plot(season1)
        plt.legend('ABCDEF', ncol=2, loc='upper left');
```

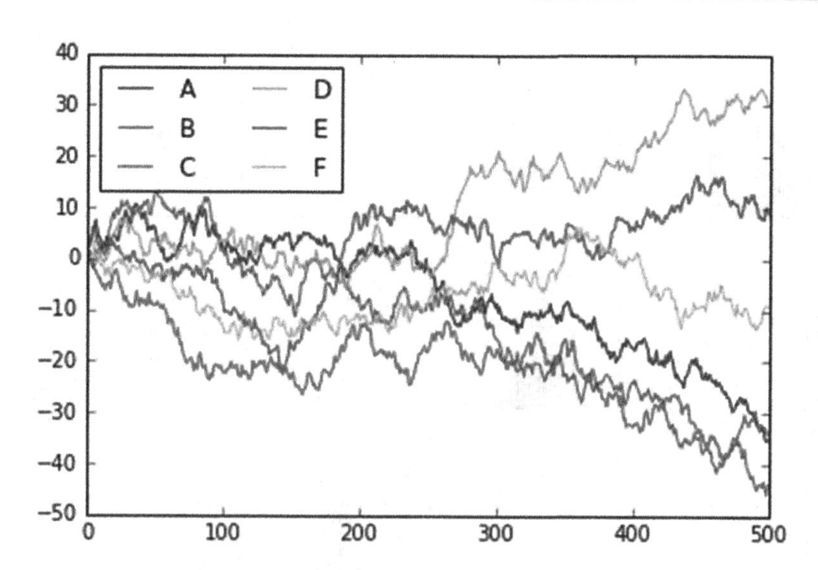

2. Load the well-known Iris data set, which lists measurements of petals and sepals of three iris species. Then plot the correlations between each pair using the `.pairplot()` method.

Answer:

See Listing 7-23.

Listing 7-23. Pair Correlations

```
In [33]: import seaborn as sns
         iris = sns.load_dataset("iris")
         iris.head()
         sns.pairplot(iris, hue='species', size=2.5);
```

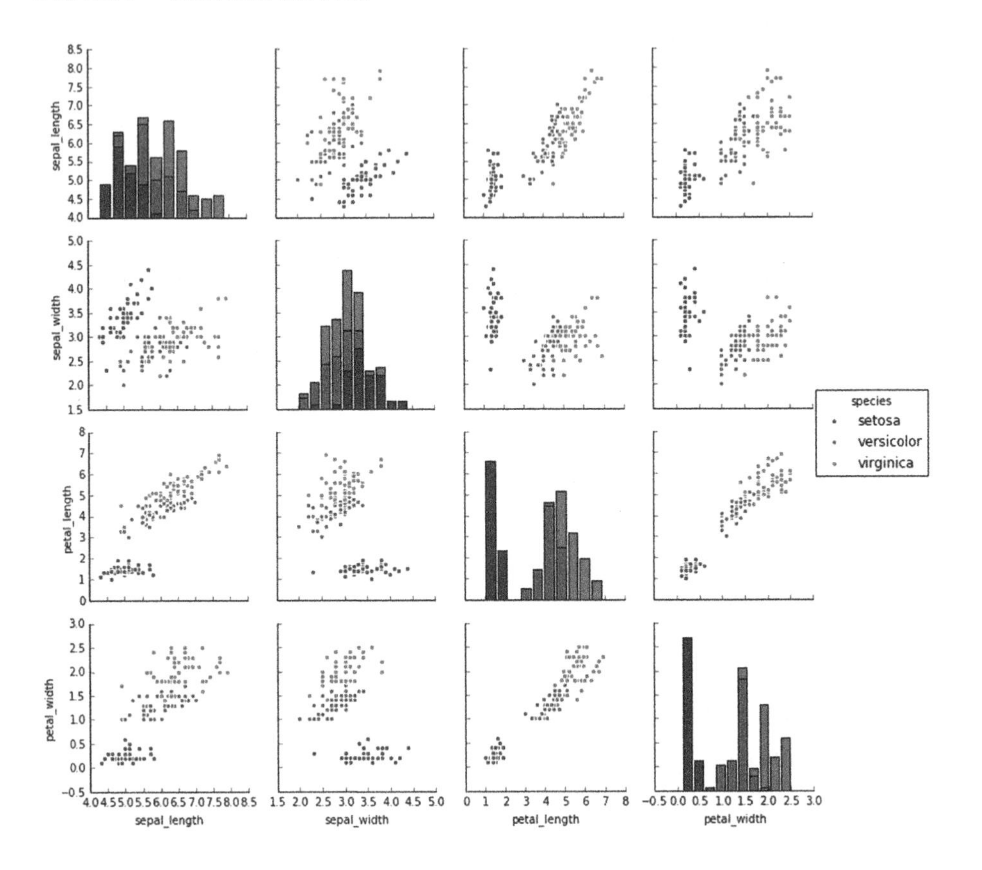

3. Load the well-known Tips data set, which shows the number of tips received by restaurant staff based on various indicator data; then plot the percentage of tips per bill according to staff gender.

Answer:

See Listing 7-24.

Listing 7-24. First five records in the Tips dataset

```
In [36]: import seaborn as sns
         tips = sns.load_dataset('tips')
         tips.head()
```

Out[36]:

	total_bill	tip	sex	smoker	day	time	size
0	16.99	1.01	Female	No	Sun	Dinner	2
1	10.34	1.66	Male	No	Sun	Dinner	3
2	21.01	3.50	Male	No	Sun	Dinner	3
3	23.68	3.31	Male	No	Sun	Dinner	2
4	24.59	3.61	Female	No	Sun	Dinner	4

```
In [37]: tips['Tips Percentage'] = 100 * tips['tip'] /
tips['total_bill']
         grid = sns.FacetGrid(tips, row="sex", col="time",
         margin_titles=True)
         grid.map(plt.hist, "Tips Percentage", bins=np.
         linspace(0, 40, 15));
```

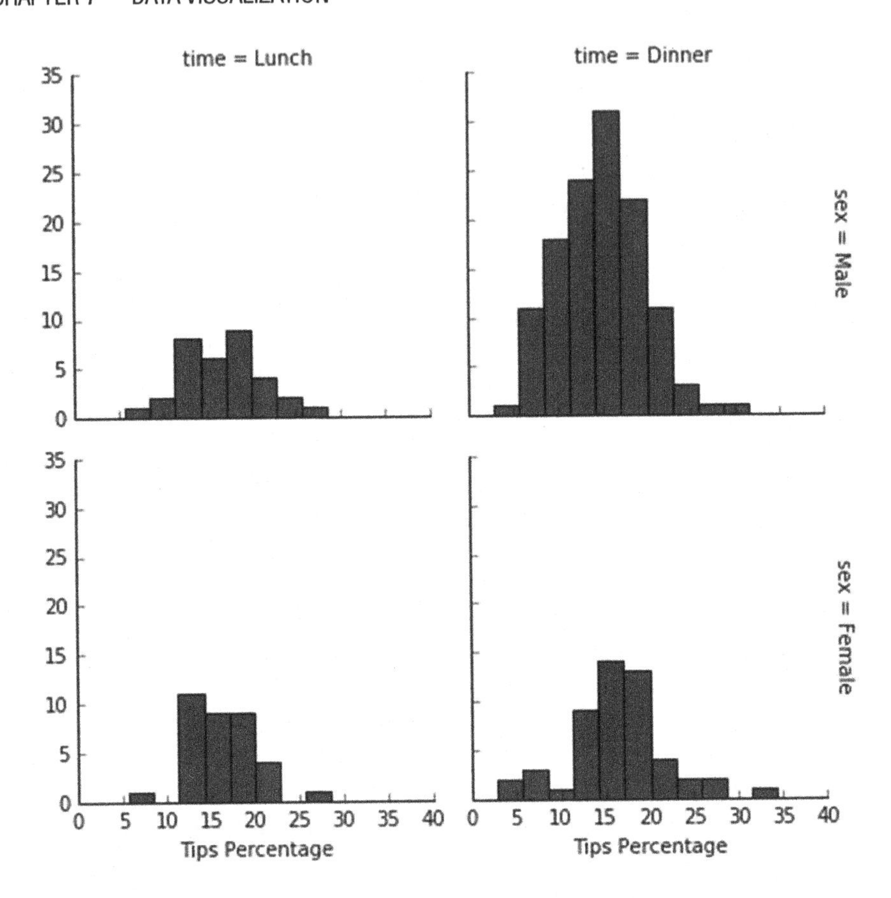

4. Load the well-known Tips data set, which shows the number of tips received by restaurant staff based on various indicator data; then implement the factor plots to visualize the total bill per day according to staff gender.

Answer:

See Listing 7-25.

Listing 7-25. Implementing Factor Plotting

```
In [39]: import seaborn as sns
         tips = sns.load_dataset('tips')
         with sns.axes_style(style='ticks'):
         g = sns.factorplot("day", "total_bill",
         "sex", data=tips, kind="box")
         g.set_axis_labels("Bill Day", "Total Bill Amount")
```

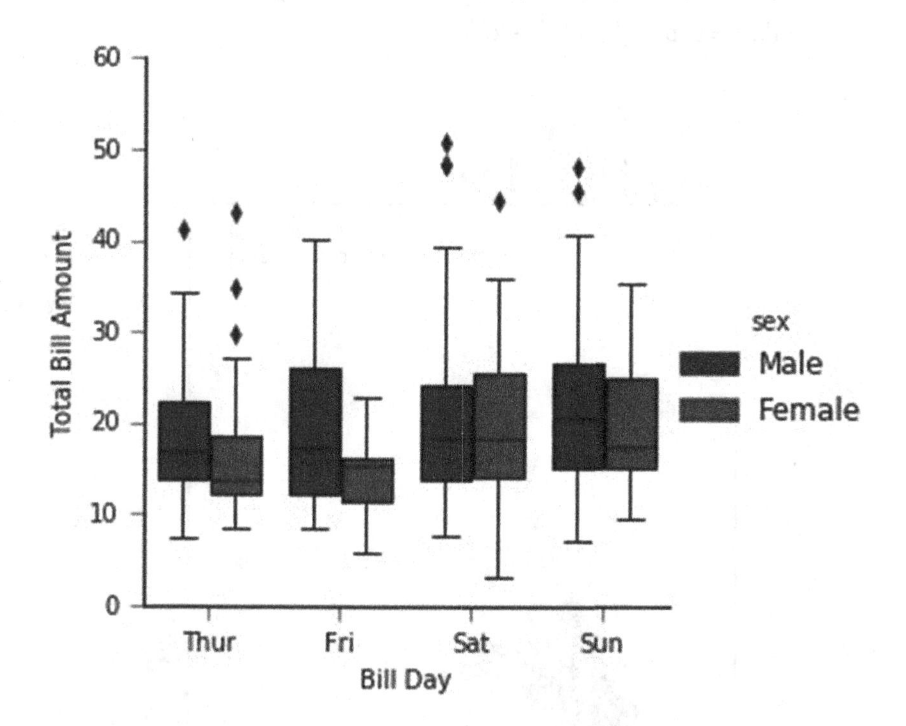

5. Reimplement the previous exercise using the Seaborn joint plot distributions.

Answer:

See Listing 7-26.

Listing 7-26. Implementing Joint Plot Distributions

```
In [43]: import seaborn as sns
         tips = sns.load_dataset('tips')
         with sns.axes_style('white'):
         sns.jointplot( "total_bill", "tip",
         data=tips, kind='hex')
```

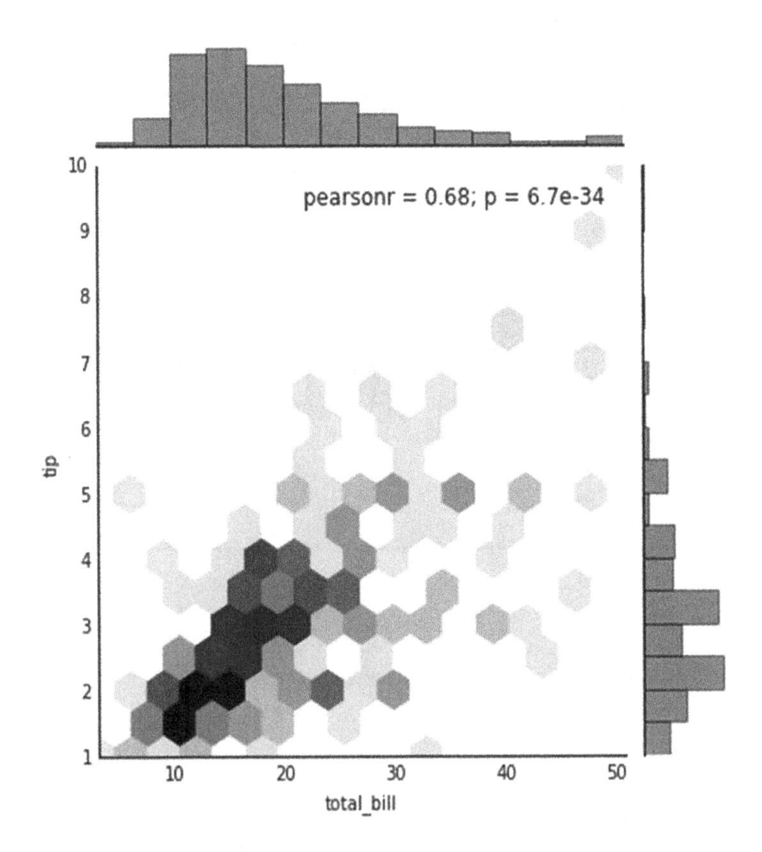

CHAPTER 8

Case Studies

This chapter covers two case studies. I will provide some brief information about each case and then show how to gather the data needed for analysis, how to analyze the data, and how to visualize the data related to specific patterns.

Case Study 1: Cause of Deaths in the United States (1999–2015)

This study analyses the leading causes of death in the United States of America between 1999 and 2015.

Data Gathering

It's important to gather a study's data set from a reliable source; it's also important to use an updated and accurate data set to get unbiased findings. The data set in this case study comes from open data from the U.S. government, which can be accessed through `https://data.gov`.

You can download it from here:

`https://catalog.data.gov/dataset/age-adjusted-death-rates-for-the-top-10-leading-causes-of-death-united-states-2013`

O. Embarak, *Data Analysis and Visualization Using Python*,
https://doi.org/10.1007/978-1-4842-4109-7_8

This case study will try to answer the following questions:

- What is the total number of records in the dataset?
- What were the causes of death in this data set?
- What was the total number of deaths in the United States from 1999 to 2015?
- What is the number of deaths per each year from 1999 to 2015?
- Which ten states had the highest number of deaths overall?
- What were the top causes of deaths in the United States during this period?

Data Analysis

Let's first read and clean the data set.

- What is the total number of recorded death cases?

See Listing 8-1.

Listing 8-1. Cleaned Records of Death Causes in the United States

```
In [2]: import pandas as pd
        data = pd.read_csv("NCHS.csv")
        data.head(3)
```

Out[2]:

	Year	113 Cause Name	Cause Name	State	Deaths	Age-adjusted Death Rate
0	1999	Accidents (unintentional injuries) (V01-X59,Y8...	Unintentional Injuries	Alabama	2313.0	52.2
1	1999	Accidents (unintentional injuries) (V01-X59,Y8...	Unintentional Injuries	Alaska	294.0	55.9
2	1999	Accidents (unintentional injuries) (V01-X59,Y8...	Unintentional Injuries	Arizona	2214.0	44.8

```
In [3]: data.shape # 15028 rows and 6 columns

Out[3]: (15028, 6)
```

Remove all rows with na cases.

```
In [4]: data = data.dropna()
        data.shape
Out[4]: (14917, 6)
```

Approximately 14,917 death cases were recorded in different U.S. states.

Now let's clean the data to find the number of death causes in the data set.

- What were the causes of death in this dataset?

See Listing 8-2.

Listing 8-2. Unique Death Causes in the United States

```
In [7]: causes = data["Cause Name"].unique()
        causes

Out[7]: array(['Unintentional Injuries', 'All Causes', "Alzheimer's disease",
               'Homicide', 'Stroke', 'Chronic liver disease and cirrhosis',
               'CLRD', 'Diabetes', 'Diseases of Heart',
               'Essential hypertension and hypertensive renal disease',
               'Influenza and pneumonia', 'Cancer', 'Suicide', 'Kidney Disease',
               "Parkinson's disease", 'Pneumonitis due to solids and liquids',
               'Septicemia'], dtype=object)
```

Remove `All Causes` from the `Cause Name` column.

```
In [8]: data = data[data["Cause Name"] !="All Causes"]
        causes = data["Cause Name"].unique()
        causes
```

```
Out[8]: array(['Unintentional Injuries', "Alzheimer's disease", 'Homicide',
               'Stroke', 'Chronic liver disease and cirrhosis', 'CLRD',
               'Diabetes', 'Diseases of Heart',
               'Essential hypertension and hypertensive renal disease',
               'Influenza and pneumonia', 'Cancer', 'Suicide', 'Kidney Disease',
               "Parkinson's disease", 'Pneumonitis due to solids and liquids',
               'Septicemia'], dtype=object)
```

```
In [9]: len(causes)
```

```
Out[9]: 16
```

As shown, there are 16 death causes according to the loaded data set. Clean the data to find the unique states included in the study. See Listing 8-3.

Listing 8-3. Unique States in the Study

```
In [11]: state = data["State"].unique()
         state
```

```
Out[11]: array(['Alabama', 'Alaska', 'Arizona', 'Arkansas', 'California',
                'Colorado', 'Connecticut', 'Delaware', 'District of Columbia',
                'Florida', 'Georgia', 'Hawaii', 'Idaho', 'Illinois', 'Indiana',
                'Iowa', 'Kansas', 'Kentucky', 'Louisiana', 'Maine', 'Maryland',
                'Massachusetts', 'Michigan', 'Minnesota', 'Mississippi',
                'Missouri', 'Montana', 'Nebraska', 'Nevada', 'New Hampshire',
                'New Jersey', 'New Mexico', 'New York', 'North Carolina',
                'North Dakota', 'Ohio', 'Oklahoma', 'Oregon', 'Pennsylvania',
                'Rhode Island', 'South Carolina', 'South Dakota', 'Tennessee',
                'Texas', 'United States', 'Utah', 'Vermont', 'Virginia',
                'Washington', 'West Virginia', 'Wisconsin', 'Wyoming'],
               dtype=object)
```

```
In [12]: data1 = data[data["State"] !="United States"]
         state = data1["State"].unique()
         state
```

```
Out[12]: array(['Alabama', 'Alaska', 'Arizona', 'Arkansas', 'California',
               'Colorado', 'Connecticut', 'Delaware', 'District of Columbia',
               'Florida', 'Georgia', 'Hawaii', 'Idaho', 'Illinois', 'Indiana',
               'Iowa', 'Kansas', 'Kentucky', 'Louisiana', 'Maine', 'Maryland',
               'Massachusetts', 'Michigan', 'Minnesota', 'Mississippi',
               'Missouri', 'Montana', 'Nebraska', 'Nevada', 'New Hampshire',
               'New Jersey', 'New Mexico', 'New York', 'North Carolina',
               'North Dakota', 'Ohio', 'Oklahoma', 'Oregon', 'Pennsylvania',
               'Rhode Island', 'South Carolina', 'South Dakota', 'Tennessee',
               'Texas', 'Utah', 'Vermont', 'Virginia', 'Washington',
               'West Virginia', 'Wisconsin', 'Wyoming'], dtype=object)
```

```
In [13]: len(state)
```

```
Out[13]: 51
```

There are 51 states included in the study.

- What was the total number of deaths in the United States from 1999 to 2015?

  ```
  In [15]: data["Deaths"].sum()
  ```

  ```
  Out[15]: 69279057.0
  ```

  The total number of deaths during the given period is 69,279,057.

- What is the number of deaths for each year from 1999 to 2015?

See Listing 8-4.

Listing 8-4. Study's Death Trends per Year

```
In [16]: dyear= data.groupby(["Year"]).sum()
         dyear
```

	Deaths	Age-adjusted Death Rate
Year		
1999	4052876.0	38550.3
2000	4054097.0	38136.3
2001	4063971.0	37645.3
2002	4104796.0	37503.0
2003	4097245.0	36904.3
2004	3999321.0	35359.7
2005	4062908.0	35368.7
2006	3990647.0	34113.0
2007	3979212.0	33405.3
2008	4038942.0	33270.1
2009	3967369.0	32052.5
2010	4001895.0	31929.8
2011	4048145.0	31522.9
2012	4069794.0	30965.9
2013	4151064.0	30930.9
2014	4213058.0	30862.1
2015	4383717.0	31496.7

```
In [18]: dyear["Deaths"].plot(title="Death per year \n
1999-2015")
```

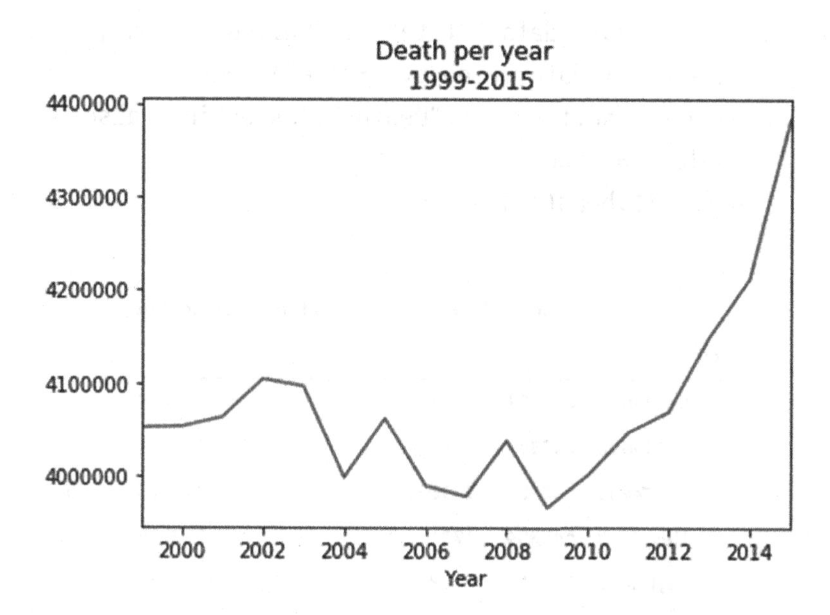

The number of deaths declined between 2002 and 2009. Then there was a continuous growth in the number of deaths from 2010 to 2013. Finally, there was a sharp increase in the number of deaths in 2013 and 2014.

Data Visualization

Plotting data gives a clear idea about patterns behind the data and helps to make the right decisions in business.

- Which ten states had the highest number of deaths overall?

See Listing 8-5.

Listing 8-5. Top Ten States with the Highest Number of Deaths in the United States

```
In [19]: data1 = data[data["State"] !="United States"]
         dataset2 = data1.groupby("State").sum()
         dataset2.sort_values("Deaths", ascending=False ,
         inplace = True)
         dataset2.head(10)
```

Out[19]:

	Year	Deaths	Age-adjusted Death Rate
State			
California	545904	3422459.0	10101.2
Florida	545904	2397507.0	10156.8
Texas	545904	2270961.0	11339.7
New York	545904	2170019.0	10226.5
Pennsylvania	545904	1785982.0	11334.1
Ohio	545904	1529552.0	11931.3
Illinois	545904	1460489.0	11170.8
Michigan	545904	1248155.0	11645.7
North Carolina	545904	1063835.0	11737.3
New Jersey	545904	1003709.0	10446.7

```
In [20]: dataset2["Deaths"].head(10).plot.bar(title= "Top ten
states with highest death number \n 1999-2015 ")
```

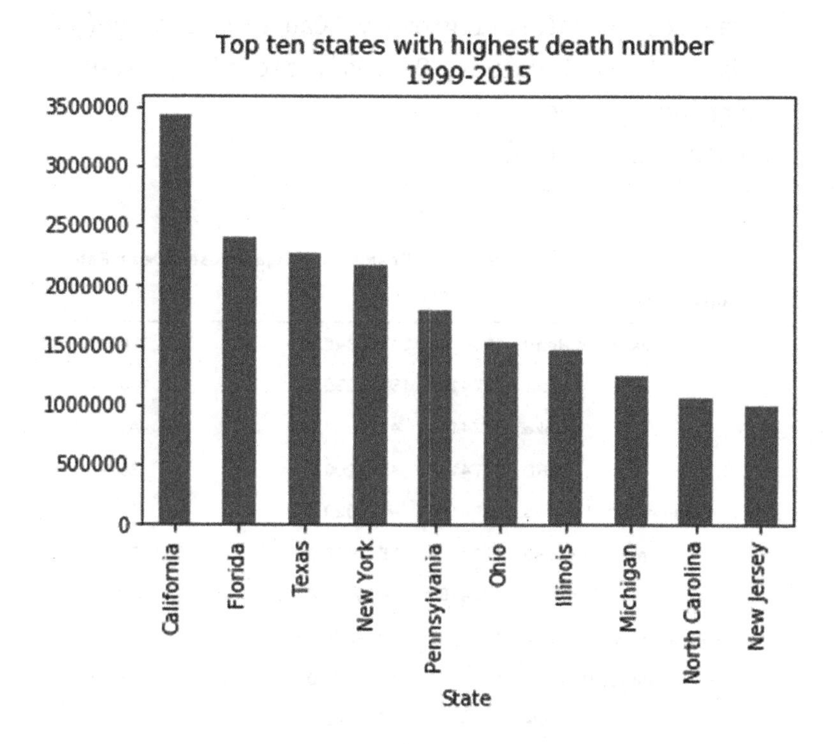

California had the highest number of deaths in the United States, with Florida coming in second.

- What were the top causes of deaths in the United States during this period?

See Listing 8-6.

Listing 8-6. Top Ten Causes of Death in the United States

```
In [21]: dataset1 = data[data["Cause Name"] !="All Causes"]
         dataset2 = dataset1.groupby("Cause Name").sum()
         dataset2.sort_values("Deaths", ascending=False ,
         inplace = True)
         dataset2.head(10)
```

Out[21]:

	Year	Deaths	Age-adjusted Death Rate
Cause Name			
Diseases of Heart	1774188	21879846.0	178315.3
Cancer	1774188	19292996.0	160163.8
Stroke	1774188	4875996.0	41458.8
CLRD	1774188	4560260.0	39545.5
Unintentional Injuries	1774188	4033020.0	37368.6
Alzheimer's disease	1774188	2514618.0	21435.6
Diabetes	1774188	2472642.0	20851.9
Influenza and pneumonia	1774188	1974864.0	16498.5
Kidney Disease	1774188	1515868.0	12555.4
Suicide	1774188	1209756.0	11580.1

```
In [22]: dataset2["Deaths"].head(10).plot.bar(title="Top ten
causes of death in USA \n 1999-2015 ")
```

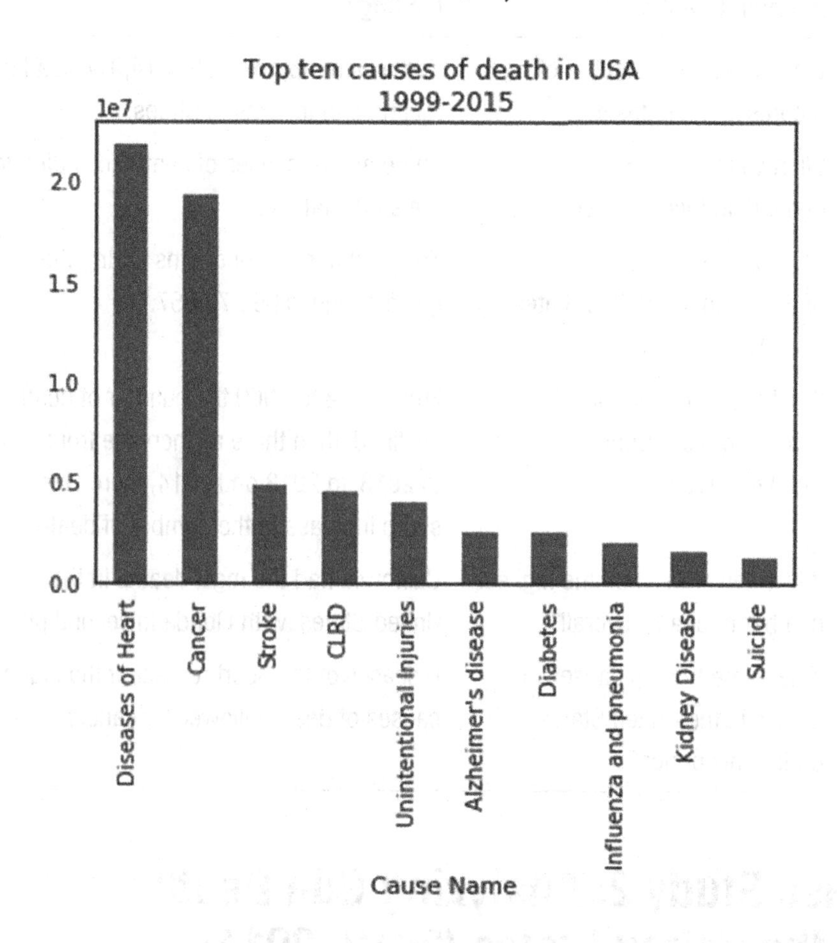

Diseases of the heart represent the biggest cause of death followed by cancer.

Findings

Table 8-1 summarizes the study findings.

Table 8-1. Case Study 1: Findings

Investigation Question	Findings
1. What is the total number of records in the dataset?	There were approximately 14,917 deaths recorded in the United States.
2. What were the causes of death in this data set?	There are 16 causes of death according to the study data set.
3. What was the total number of deaths in the United States from 1999 to 2015?	The total number of deaths during the given period is 69,279,057.
4. What is the number of deaths per year from 1999 to 2015?	From 2002 to 2009 the number of deaths declined, then there an increase from 2010 to 2013. In 2013 and 2014, there was a sharp increase in the number of deaths.
5. Which ten states had the highest number of deaths overall?	California had the most deaths in the United States, with Florida in second place.
6. What were the top causes of deaths in the United States during this period?	Diseases of the heart represent the highest causes of death followed by cancer.

Case Study 2: Analyzing Gun Deaths in the United States (2012–2014)

This study analyzes gun deaths in the United States of America between 2012 and 2014.

This case study will try to answer the following questions:

- What is the number of annual suicide gun deaths in the United States from 2012 to 2014, by gender?

- What is the number of gun deaths by race in the United States per 100,000 people from 2012 to 2014?
- What is the annual number of gun deaths in the United States on average from 2012 to 2014, by cause?
- What is the percentage per 100,000 people of annual gun deaths in the United States from 2012 to 2014, by cause?
- What is the percentage of annual suicide gun deaths in the United States from 2012 to 2014, by year?

Data Gathering

The data set for this study comes from GitHub and can be accessed here:

```
https://github.com/fivethirtyeight/guns-data.git
```

Load and clean the dataset and prepare it for processing. See Listing 8-7.

Listing 8-7. Reading Gun Deaths in the United States (2012–2014) Data Set

```
In [25]: import pandas as pd
         import numpy as np
         import matplotlib.pyplot as plt
         import seaborn as sns
         sns.set(style='white', color_codes=True)
         %matplotlib inline
```

```
In [26]: dataset = pd.read_csv('Death data.csv', index_col=0)
         print(dataset.shape)
         dataset.index.name = 'Index'
         dataset.columns = map(str.capitalize, dataset.columns)
         dataset.head(5)

         (100798, 10)
```

Out[26]:

Index	Year	Month	Intent	Police	Sex	Age	Race	Hispanic	Place	Education
1	2012	1	Suicide	0	M	34.0	Asian/Pacific Islander	100	Home	BA+
2	2012	1	Suicide	0	F	21.0	White	100	Street	Some college
3	2012	1	Suicide	0	M	60.0	White	100	Other specified	BA+
4	2012	2	Suicide	0	M	64.0	White	100	Home	BA+
5	2012	2	Suicide	0	M	31.0	White	100	Other specified	HS/GED

Organize the data set by year and then by month.

```
In [27]: dataset_Gun = dataset
          dataset_Gun.sort_values(['Year', 'Month'],
          inplace=True)
```

Data Analysis

Now let's look at the data and make some analysis.

- How many males and females are included in this study?

```
In [28]: dataset_Gun.Sex.value_counts(normalize=False)
Out[28]: M    86349
F    14449
Name: Sex, dtype: int64
```

- How many educated females are included in this study?

  As shown here, there are 14,243 educated females involved in this study.

  Group the data set by gender.

```
In [8]: dataset_byGender = dataset_Gun.groupby('Sex').
count()
dataset_byGender
```

Out[8]:

Sex	Year	Month	Intent	Police	Age	Race	Hispanic	Place	Education
F	14449	14449	14449	14449	14446	14449	14449	14386	14243
M	86349	86349	86348	86349	86334	86349	86349	85028	85133

Data Visualization

In this case study, we will try to find the answers to the numerous questions posed earlier. Let's get started.

- What is the number of suicide gun deaths in the United States from 2012 to 2014, by gender?

See Listing 8-8.

Listing 8-8. Gun Death by Gender

```
In [29]: dataset_suicide_Gender =dataset_Gun[
         dataset_Gun["Intent"] =="Suicide"]
         dataset_suicide_Gender.Sex.value_counts
         (normalize=False).plot.bar(title='Annual U.S.\\suicide
         gun deaths \n 2012-2014, by gender')
```

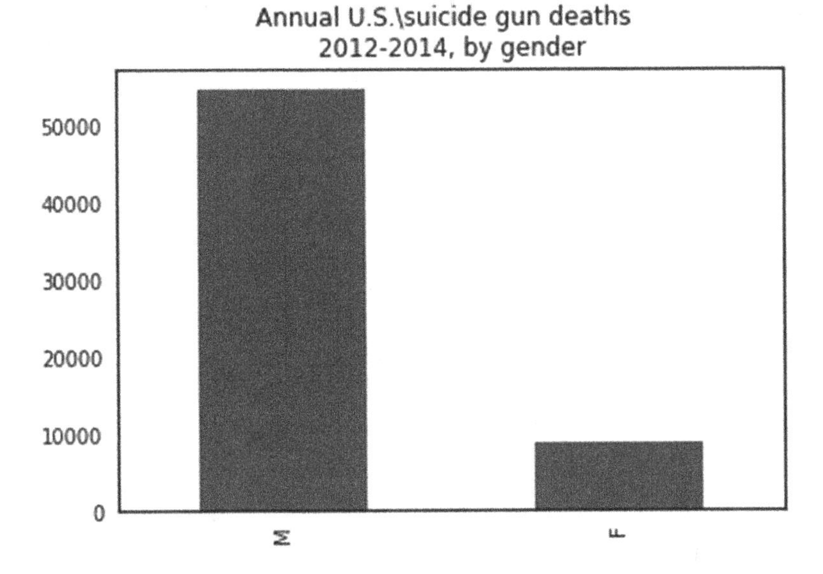

It's clear that there are huge differences between males and females. The number of male suicides by gun is above 50,000, while the female death rate is below 10,000, which shows how males are more likely to commit suicide using a gun.

```
In [31]: dataset_byGender.plot.bar(title='Annual U.S. suicide
gun deaths \n 2012-2014, by gender')
```

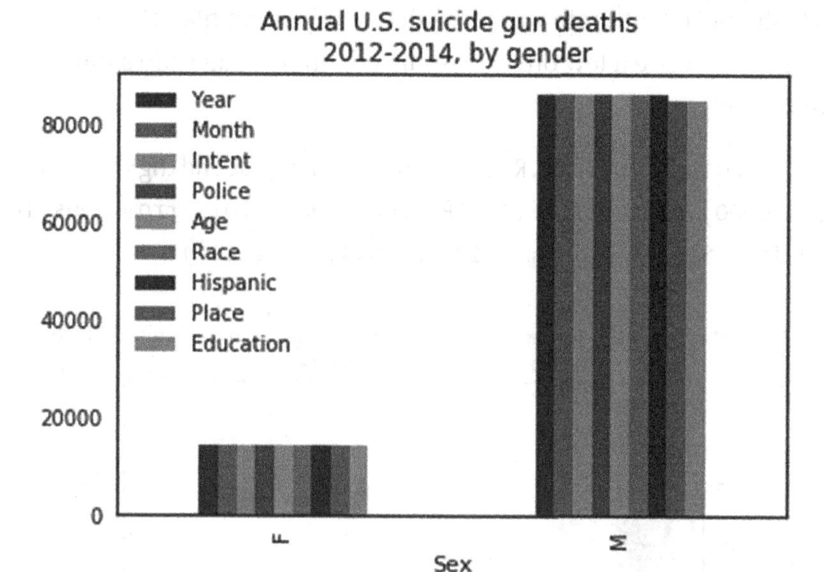

- What is the number of gun deaths by race in the United States per 100,000 people from 2012 to 2014?

See Listing 8-9.

Listing 8-9. Analyzing and Visualizing Gun Death Percentage by Race

```
In [32]: dataset_byRace = dataset (dataset_byRace.Race.value_
counts(ascending=False)*100/100000)
```

```
Out[32]: White                                  66.237
         Black                                  23.296
         Hispanic                                9.022
         Asian/Pacific Islander                  1.326
         Native American/Native Alaskan          0.917
         Name: Race, dtype: float64
```

The highest death rate was for white people, then black, and then Hispanic. There are a few other races listed, but the rates are small comparatively.

```
In [33]:(dataset_byRace.Race.value_counts(ascending=False)
*100/100000).plot.bar(title='Percent death toll from guns in
the United States \nfrom 2012 to 2014, by race')
```

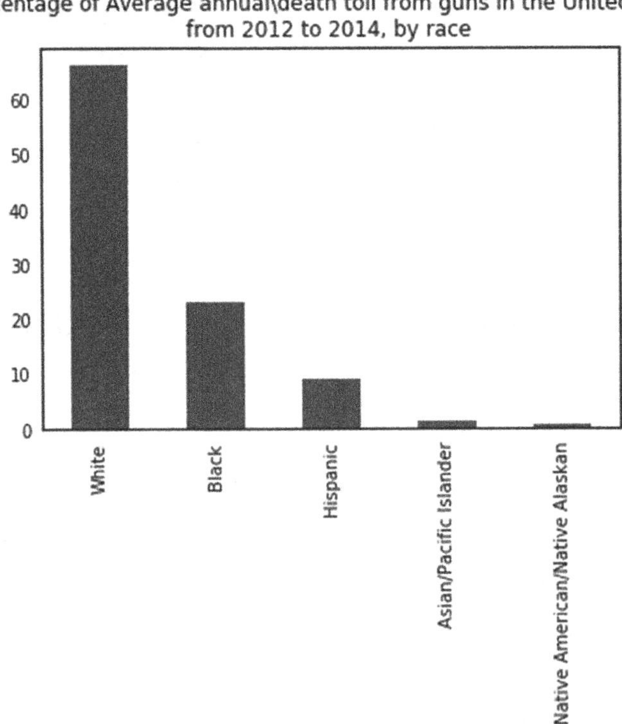

- What is the number of gun deaths in the United States on average from 2012 to 2014, by cause?

See Listing 8-10.

Listing 8-10. Visualizing Gun Death by Cause

```
In [14]: dataset_byRace.Intent.value_counts(sort =True,
ascending=False)
```

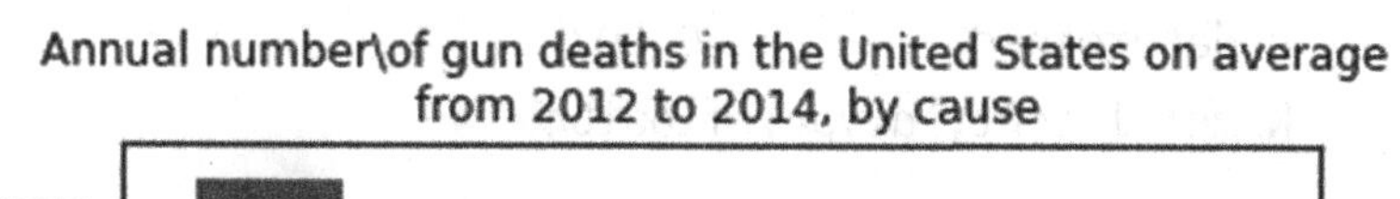

```
Out[14]: Suicide         63175
         Homicide        35176
         Accidental       1639
         Undetermined      807
         Name: Intent, dtype: int64
```

```
In [17]: dataset_byRace.Intent.value_counts(sort=True).plot.
bar(title='Annual number of gun deaths in the United States on
average \n from 2012 to 2014, by cause')
```

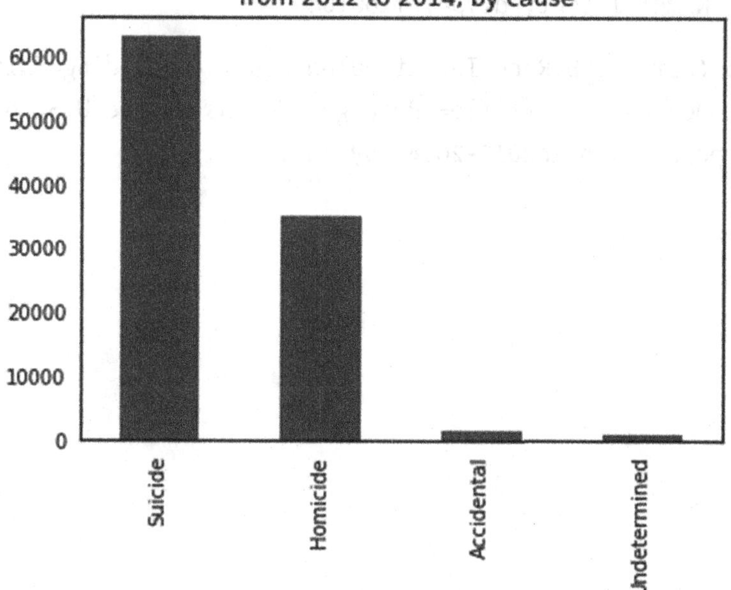

The figure shows a high number of suicide and homicide deaths compared to a low number of deaths due to accidents.

- What is the percentage per 100,000 people of annual gun deaths in the United States from 2012 to 2014, by cause?

See Listing 8-11.

Listing 8-11. Visualizing Gun Death per 100,000 by Cause

```
In [40]: dataset_byRace.Intent.value_counts(ascending=False)
*100/100000

Out[40]: Suicide               63.175
         Homicide              35.176
         Accidental             1.639
         Undetermined           0.807
         Name: Intent, dtype: float64

In [41]: (dataset_byRace.Intent.value_counts(ascending=False)
*100/100000).plot.bar(title='Rate gun deaths in the U.S. per
100,000 population \n2012-2014, by race')
```

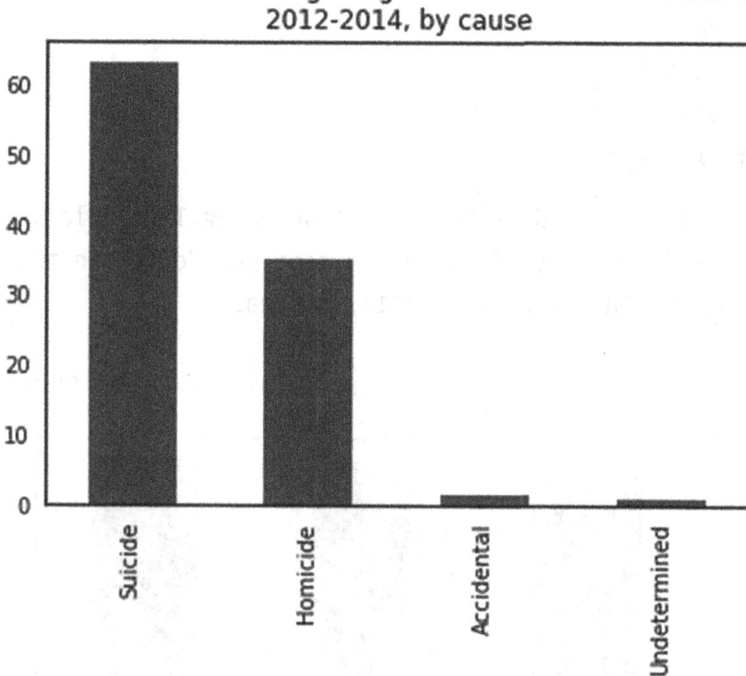

This shows that there are 60 suicide cases for every 100,000 people. In addition, there are 30 homicide cases for every 100,000.

- What is the percentage of suicide gun deaths in the United States from 2012 to 2014, by year?

See Listing 8-12.

Listing 8-12. Visualizing Gun Death by Year

```
In [42]: dataset_suicide=dataset[ dataset["Intent"]
=="Suicide"]
datasetSuicide= dataset_suicide.Year.value_
counts(ascending=False) *100/100000
datasetSuicide.sort_values(ascending=True)
```

```
Out[42]:
2012    20.666
2013    21.175
2014    21.334
Name: Year, dtype: float64
```

```
In [43]:datasetSuicide.sort_values(ascending=True).plot.
bar(title='Percentage of annual suicide gun deaths in the
United States \nfrom 2012 to 2014, by year')
```

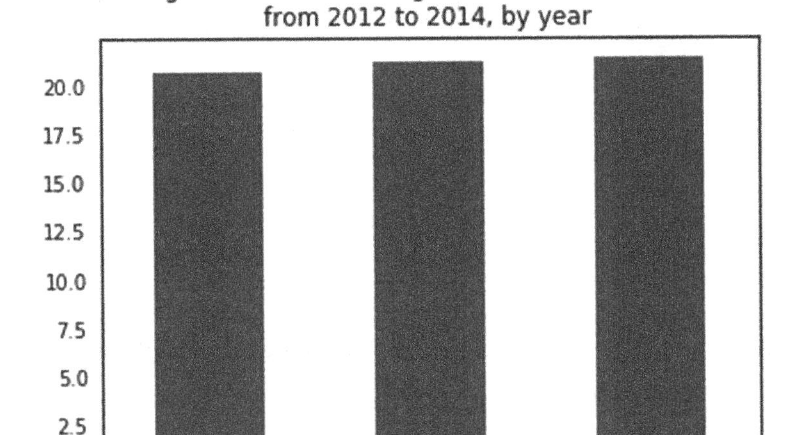

The figure shows almost the same number of suicides each year over three years, which means that this is a regular pattern.

Findings

Table 8-2 shows the findings.

Table 8-2. *Case Study 2: Findings*

Investigation Question	Findings
1. What is the number of U.S. suicide gun deaths from 2012 to 2014, by gender?	Male suicide gun deaths is over 50,000, while females suicide gun deaths is below 10,000, which shows how males are more likely to commit suicide with a gun.
2. What is the number of gun deaths in the United States per a 100,000 population from 2012 to 2014?	The highest number of deaths is for while people, then black, and then Hispanic.
3. What are the annual number of gun deaths in the United States on average from 2012 to 2014, by cause?	There is a high number of suicide and homicide deaths compared to a low number of deaths due to accidents.
4. What is the 100,000 percentage of annual guns death tolls in the United States from 2012 to 2014, by cause?	The 100,000 percentages shows that there are 60 suicide cases for every 100,000 people, which somehow is not a high rate. In addition, there are 30 homicide cases for every 100,000 people.
5. What is the percentage of annual suicide gun deaths in the United States from 2012 to 2014, by year?	The analysis shows almost the same number of suicides each year over a period of three years, which means that this is a regular pattern in society.

Summary

This chapter covered how to apply Python techniques on two different case studies. Here's what you learned:

- How to determine the problem under investigation
- How to determine the main questions to answer
- How to find a reliable data source
- How to explore the collected data to remove anomalies
- How to analyze and visualize cleaned data
- How to discuss findings

Index

O. Embarak, *Data Analysis and Visualization Using Python*,
https://doi.org/10.1007/978-1-4842-4109-7

E, F

G

H

I

J, K

L

M

N, O

P, Q

R

S

T, U, V

W

X, Y, Z

Printed in the United States
By Bookmasters